COMMUNICATION SYSTEMS ANALYSIS

P. B. JOHNS, B.Sc.(Eng.), M.Sc., C.Eng., M.I.E.E.
Lecturer in Communications, University of Nottingham

and

T. R. ROWBOTHAM, B.Sc.(Eng.), M.Sc., C.Eng., M.I.E.E.
Senior Executive Engineer, Research Dept., Post Office

LONDON BUTTERWORTHS

THE BUTTERWORTH GROUP

ENGLAND

Butterworth & Co (Publishers) Ltd
London: 88 Kingsway WC2B 6AB

AUSTRALIA

Butterworth & Co (Australia) Ltd
Sydney: 586 Pacific Highway Chatswood. NSW 2067
Melbourne: 343 Little Collins Street, 3000
Brisbane: 240 Queen Street, 4000

CANADA

Butterworth & Co (Canada) Ltd
Toronto: 14 Curity Avenue, 374

NEW ZEALAND

Butterworth & Co (New Zealand) Ltd
Wellington: 26–28 Waring Taylor Street, 1
Auckland: 35 High Street, 1

SOUTH AFRICA

Butterworth & Co (South Africa) (Pty) Ltd
Durban: 152–154 Gale Street, 1

First published 1972

ISBN 0 408 70197 8 Standard
0 408 70198 6 Limp

Printed in Hungary

FOREWORD

The general approach in this book, with its major emphasis on fundamental principles, but with sufficient background of communication systems to make the application of the principles convincingly intelligible, is one that should enable it to achieve an important place in University teaching at under-graduate level. It should also prove of value in the training of PO and industry staff concerned with telecommunications.

The inclusion of a range of illustrative examples, in question and answer form, and representative of the problems encountered in system design is an excellent feature.

W. J. BRAY
Director of Research, Post Office

November 1970

PREFACE

This undergraduate level book is orientated to the practical problems encountered in communication engineering. Interference between radio systems, signal/noise calculations for satellite communications, the effect of nonlinearites on signals, and pulse code modulation system design are typical examples of items of current interest which are covered.

An extremely simple explanation of the basic properties of noise and realistic signals (multiplexed telephony in particular) is presented to the designer in Chapter 1. Subsequent chapters deal with the application of these basic mathematical tools to a range of modern communication problems and clarification is assisted by the liberal use of worked and unworked examples.

As demand for communication channels increases, systems are becoming more sophisticated and their design can require a working knowledge of quite a wide range of specialised literature in the subject. This book provides an introduction to this literature and also attempts to compile some of the more important results useful to the designer. For example, formulae are tabulated for the thermal noise performance of the commonly used analogue and digital systems, and the effect of nonlinearities on signal/noise ratios are given for a comprehensive list of nonlinearity curves.

The practical nature of this book not only provides a basis for a new approach to communications in second and third year university undergraduate courses but also the results derived make it a useful handbook for the systems design engineer.

P.B.J. T.R.R.

CONTENTS

Chapter 1

INTRODUCTION AND BASIC MATHEMATICS

1.1. INTRODUCTION

The enormous increase in communication requirements over recent years has led to the development of means of transmitting more and more information simultaneously between two points. The growth in demand arises from many sources, both old and new, such as the convenience of telephoning and the distribution of television both on national and international scales. Broadcast systems are expanding particularly in the v.h.f. (very high frequency) frequency bands, and soon computers will be passing vast amounts of data from one centre to another.

One aspect of the problem of carrying large amounts of information from one point to another over a single communication link is in providing channels capable of passing high enough frequencies i.e. channels of large enough bandwidth. For example, a television signal requires signal frequencies up to several MHz and before the era of communication satellites transoceanic television pictures were impossible simply because there was not enough bandwidth on submarine cables. In order to transmit information requiring large bandwidths therefore, the frequencies on channels carrying the information have had to become progressively higher.

A number of the systems which allow these large bandwidths to be transmitted will be considered in the following paragraphs together with the types of signal used to transmit information. The way in which the transmission of these signals through 'noisy' transmission paths affects the communication system forms the basis of the subsequent chapters of the book.

Radio relay systems using frequency modulation and operating in the 4 GHz and 6 GHz bands are currently in use, and such systems may carry numbers of radio channels each consisting of as many as 2 700 channels of telephony with modulating signals of the order of 12 MHz. Microwave systems of this type can be built with highly directive aerials thus concentrating the transmitted energy into relatively narrow beams, the propagation being essentially in straight lines. Transmitting and receiving aerials must be approximately in optical line of sight, and radio towers carrying microwave dish or horn aerials are becoming almost commonplace. Circuits may consist of many stations, each some 30 miles apart, relaying large amounts of information from tower to tower and thereby around the curvature of the earth.

Coaxial cable with repeaters have of course been used for a long time to carry trunk telephony and television and a single route can consist of a number of separate cables each again carrying as many as 2 700 telephone channels.

Much consideration is being given to transmitting information at microwave frequencies along waveguides and in particular to using circular waveguides and employing a low loss waveguide mode. Pulse code modulation is likely to be used for these systems and operating frequencies may be as high as 100 GHz enabling enormous amounts of information to be carried.

Communication satellites which also use microwave frequencies have developed very quickly and the modern concept of many countries using a single satellite seems very remote from the first experiments when signals were bounced off passive balloons. Total capacities (to be shared among a number of countries) of 4 000 telephone channels are possible and the satellites (which are in the synchronous orbit) enable most countries to communicate with each other. Frequency modulation is used for these systems but digital modulation methods are likely to become increasingly important in the future.

Although submarine cables have considerably less capacity than satellite systems, they are used to great advantage where a smaller number of permanent circuits are required. They also have the advantage of greater security over satellite systems.

Transmission frequencies of broadcast systems have also tended to increase. In order to provide high quality sound, free from interference, and also to provide stereo transmissions, the v.h.f. band is used for sound broadcasting (about 100 MHz). The use of 625 line television for monochrome and colour has led to domestic reception at frequencies well into the u.h.f. (ultra high frequency) bands (600 MHz approximately).

Many of the design techniques dealt with in the following chapters of this book are common to all of these communication systems. Thermal noise, for example, is present in any communication system and analysis techniques for this form of noise are given in Chapter 2.

Digital modulation methods are becoming increasingly important both in the trunk communication systems described above (pulse code modulation) and as the basis of research into more specialised communication applications such as bandwidth compression (delta modulation) and pulse compression (chirp modulation). Noise is inherent to these modulation methods and the basic theory for digital system design appears in Chapter 6. The effect of thermal noise on the reception of digital signals is also covered in Chapter 6.

The fundamentals for the analysis of systems using analogue modulation methods (frequency, phase, and amplitude modulation) are presented in Chapter 3 together with computational methods for obtaining their spectra. Since line-of-sight radio relay systems and communication satellite systems share the same frequency bands interference problems can arise between these systems and so special emphasis is placed on this aspect of system design.

Noise due to distortion from nonlinearities is present in most communication systems and Chapter 5 provides the design theory for this type of problem. Distortion of multiple signals due to nonlinearity such as that found in the output travelling wave tube in a multiple carrier satellite is an example of a particularly important design problem. Special attention is paid to this type of distortion in Chapter 5.

Distortion noise due to echoes in frequency modulated systems, such as that encountered in the feeders of line-of-sight radio relay systems, is covered in Chapter 4. This Chapter also deals with amplitude modulation to phase modulation (a.m. to p.m.) conversion in devices. This problem again arises for example in the output travelling wave tube of a satellite.

These Chapters (2 to 6) are intended to be independent of each other as much as possible to enable study of any particular design problem without reference to other chapters. The basic mathematical background needed for all of these chapters is kept to a minimum and this, together with an introduction to the basic modulation methods, occupies the remainder of this chapter.

1.2. OPERATIONS ON SIGNALS

1.2.1. Multiplexing

In many of the communication systems mentioned in the previous sections, signals are required to share the same communication channel. For example in telephony systems it is possible for many telephone conversations to take place simultaneously over a single pair of wires.

Sharing or multiplexing is achieved by either dividing up the frequency spectrum or by dividing up time. In the former method telephone channels (for example) are stacked on top of each other in frequency. For instance if a telephone channel occupies a bandwidth of 4 KHz then one channel can occupy the frequencies 0–4 kHz. A second channel may then be translated up in frequency to occupy a band 4 kHz–8 kHz and the two channels transmitted simultaneously in a bandwidth of 0–8 kHz. This method of multiplexing is termed frequency division multiplex (f.d.m.). Time division multiplexing (t.d.m.) is achieved by sampling the information, e.g. by noting the amplitude at equal intervals of time, on one channel and then the next and so on, and transmitting each sample in turn. Sampling is considered in more detail in Chapter 6 while some of the problems arising from sharing within communication systems is discussed further in Chapter 3.

It follows from the above that the bandwidth of n channels in f.d.m. is n times the bandwidth of one channel. Also for a given sampling rate of a channel, the overall sampling rate for n channels in t.d.m. is n times the sampling rate of one channel. It is shown in Chapter 6 that the whole of the information of a channel which is sampled at a sufficiently fast rate is contained in a bandwidth equal to the channel itself, and therefore it can be concluded that in theory the bandwidth of a communication channel required for n channels in f.d.m. is the same as that required for n channels in t.d.m.

1.2.2. Modulation

In many cases the information signal is in an unsuitable form for transmission between one point and another and some kind of translation (to a higher frequency, for example) must take place. This process is termed modulation and the retranslation back to the original signal is termed demodulation.

There are two main classes of modulation – analogue modulation and digital modulation, and two main types of waveform are used to carry the modulation – continuous sine wave carrier and a pulse stream.

1.2.3. Analogue Modulation

Analogue modulation is the process of varying (modulating) a parameter of a waveform in sympathy with an information waveform. As indicated the waveform to be modulated (the carrier) is usually either sinusoidal or a pulse stream and the common analogue modulation methods are mentioned for each in turn in this section.

Consider the sine wave carrier,

$$A \sin (\omega_c t + \phi) \tag{1.1}$$

where A is the amplitude of the carrier
ω_c is the angular frequency which is considered to be large compared with the frequencies in the modulating signal
ϕ is a relative phase parameter

The modulating signal waveform $f(t)$ may be used to vary the amplitude A of the carrier (amplitude modulation), and it is usual to keep all amplitudes positive by providing $f(t)$ with a d.c. bias. Alternatively the waveform $f(t)$ may be used to vary the phase ϕ (frequency or phase modulation). The way in which the signal information is spread about the carrier frequency (the spectrum) for these types of modulation are considered in detail in Chapter 3.

The modulating signal waveform $f(t)$ may also be used to vary the heights of pulses in a pulse stream, or the width of pulses in a pulse stream, and an analysis of this type of modulation is given in Chapter 6.

In all these modulation methods the precise variation of the signal waveform $f(t)$ is used to vary a parameter on the carrier and because the variation is a continuum the modulation is referred to as analogue.

1.2.4. Digital Modulation

In this form of modulation only a fixed number of defined and discrete amplitudes can be transmitted from one point to another. The modulating signal $f(t)$ must first be processed in order to define it in terms of the fixed number of amplitudes and this processing inevitably introduces noise. The advantage of this type of modulation, which is also considered in detail in Chapter 6, is that the receiver can have prior knowledge of the fixed amplitudes and need only make a decision as to which amplitude is being transmitted at a certain time.

Thus although a transmission path may give rise to distortion, a receiver which is able merely to decide which type of signal is being sent, regardless of how much the signal shape is distorted, will regenerate the original signal perfectly. Regeneration of an analogue signal in this way is impossible and the distortion appears as noise. Completely random noise (thermal for example) on a transmission path will cause errors on a digital system and these will be cumulative in the same way as the noise is cumulative for the analogue system. Digital systems, however, can be much more immune to certain types of noise than analogue systems. This is exemplified by the fact that 24 channels of digitally modulated telephony may be transmitted between telephone exchanges on wires originally only capable of taking one channel. The increased efficiency arises because the modulation used (pulse code modulation) is less susceptible to noise and crosstalk in wires in multi-cored telephone cables.

1.3. BASIC MATHEMATICAL CONCEPTS FOR NOISE ANALYSIS

1.3.1. Probability Distribution

Noise is inherent to any communication system, and the problem is to design the communication system for adequate signal/noise ratio at the output of the receiver. Noise arising in communication systems takes many forms, some of which are natural like thermal noise due to molecular agitation within materials. Distortion also gives rise to loss of information. As it is usually not practical to compensate for distortion characteristics, which might depend on component tolerances for example, the distortion waveform is not known exactly. Thus for the purpose of analysis, the resultant distortion waveform can be treated as noise.

Noise can also be man made, for example interference from electrical machinery or from radio transmitters in overcrowded frequency bands.

Noise is unpredictable; if it were predictable it could be cancelled out. Information is also unpredictable; if it were predictable it would not be information. For these reasons signals and noise can be very alike in character and in fact for much of this book the signal will be treated statistically in exactly the same way as noise.

Since noise-like waveforms are unpredictable it is impossible to write down a function defining them completely, but rather they must be expressed in terms of probability functions. Figure 1.1 shows a noise-like waveform and it can be seen that the amplitude of the waveform spends more of its time around zero amplitude than it does at extreme positive or negative amplitudes. Thus it is more probable that subsequent amplitudes will be around the zero mark rather than at extreme amplitudes. Mathematically the

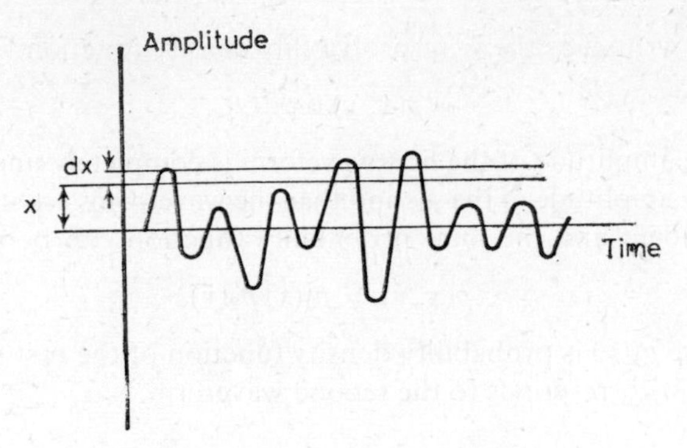

Figure 1.1 Random waveform

probability can be written as:

$$p(x)\,\mathrm{d}x = \text{The probability of finding the waveform within the amplitudes } x \text{ and } (x+\mathrm{d}x). \tag{1.2}$$

$p(x)$ is a density function, a typical form being shown in Figure 1.2. For this particular example the probability of the random variable being exactly at a certain amplitude is zero, the probability that the amplitude lies between the values α and β is given by

$$\int_{\alpha}^{\beta} p(x)\,\mathrm{d}x \tag{1.3}$$

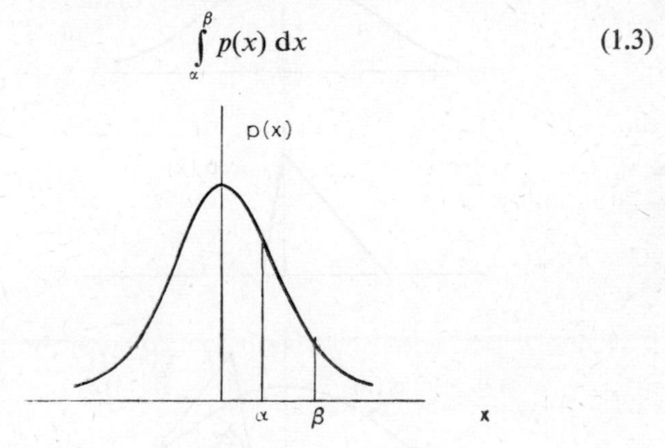

Figure 1.2 Typical probability distribution

Since the amplitude of the waveform must lie somewhere between $-\infty$ and $+\infty$

$$\int_{-\infty}^{\infty} p(x)\,\mathrm{d}x = 1 \tag{1.4}$$

If there are two random variables, the probability that one waveform is within the amplitudes x and $(x+\mathrm{d}x)$ and at the same time the amplitude of the second waveform is between y and $(y+\mathrm{d}y)$

may be written as the joint probability density function

$$p(x, y)\,\mathrm{d}x\,\mathrm{d}y \tag{1.5}$$

If the amplitude of the first waveform is completely unconnected with the amplitude of the second then the waveforms are statistically independent and the joint probability function can be written as

$$p(x, y) = p_1(x)\,p_2(y) \tag{1.6}$$

Where $p_1(x)$ is probability density function of the first waveform and $p_2(y)$ corresponds to the second waveform.

1.3.2. The Probability Distribution of the Sum of Two Random Variables

Let $p_1(x)$ be the probability density of one random variable and $p_2(x)$ be the probability density of a second which is statistically independent. If $p_{1+2}(x)$ is the probability density that the sum of the two variables is at a value x, then when the first waveform is

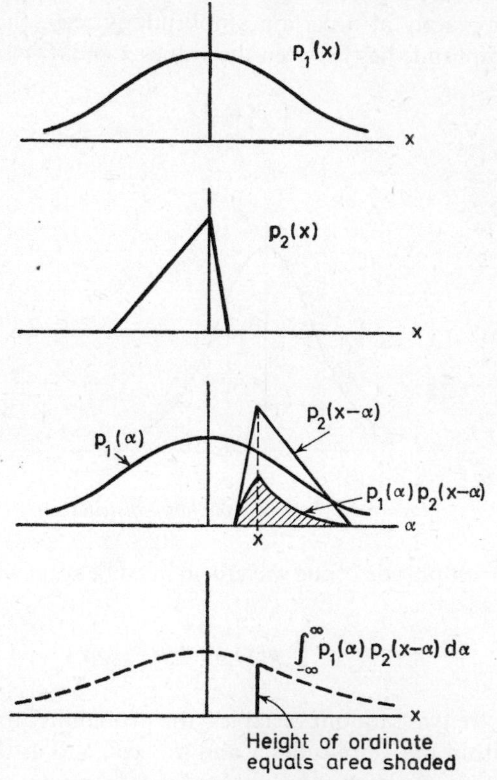

Figure 1.3 Convolution of two functions

at a value α (say) with a probability density $p_1(\alpha)$ then the second must be at a value $(x-\alpha)$ when its probability density will be $p_2(x-\alpha)$. α can be any value and the probability that the total is at a value x is the sum of all the probabilities corresponding to all the values of α

i.e.
$$p_{1+2}(x) = \int_{-\infty}^{\infty} p_1(\alpha)p_2(x-\alpha)\,d\alpha \qquad (1.7)$$

The right hand side of equation 1.7 is known as a convolution integral and is very important in communication theory in this and other applications. The value of the convolution integral is the area under the curve formed by the product of the function p_1, and the function p_2 reversed and displaced by an amount x, as shown in Figure 1.3.

Example 1.1

A random variable whose amplitude is equally likely to be at any amplitude between $-\alpha$ and $+\alpha$ and zero elsewhere is added to a waveform whose amplitude is equally likely to be at any amplitude between $-\beta$ and $+\beta$ and zero elsewhere. Calculate the probability distribution of the sum of the two waveforms.

Solution

The probability distribution of the functions are rectangles extending from $-\alpha$ to $+\alpha$, $[p_\alpha(y)]$ in the one case and from $-\beta$ to $+\beta$, $[p_\beta(y)]$ in the other; as shown in Figure 1.4. Note that the area of these rectangles is unity.

From equation 1.7, the probability distribution of the sum is the convolution of the two rectangles. Thus it is required to form the integral,

$$p_{\alpha+\beta}(x) = \int_{-\infty}^{\infty} p_\alpha(y)p_\beta(x-y)\,dy$$

For the value of x shown in Figure 1.4a, the product is $\frac{1}{4\alpha\beta}$ and the area under the product is given by

$$\text{Area} = \frac{1}{4\alpha\beta}[\alpha-(x-\beta)] = \frac{1}{4\alpha\beta}(\alpha+\beta-x)$$

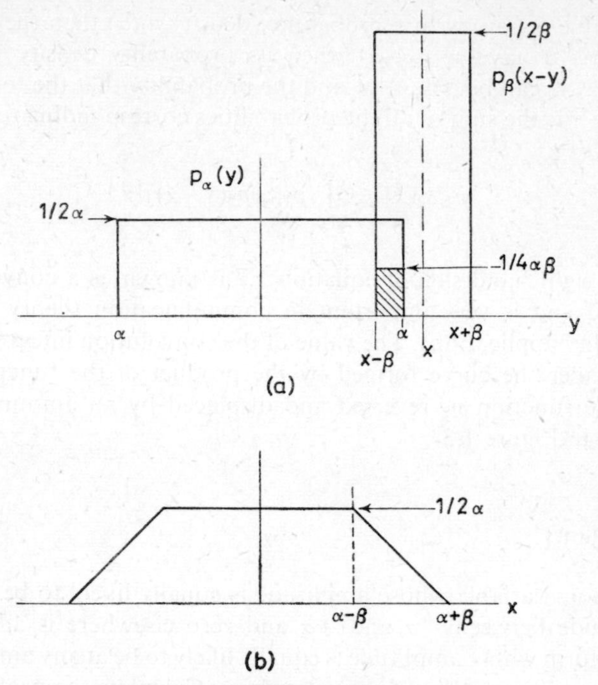

Figure 1.4 Convolution of two rectangles

The convolution is thus a ramp from $x = \alpha+\beta$ to $x = \alpha-\beta$ whereupon the convolution is constant of value

$$\frac{2\beta}{4\alpha\beta} = \frac{1}{2\alpha}$$

The process is the same for negative x and the final probability distribution is shown in Figure 1.4b. Note that the total area of the convolution is unity as would be expected.

1.3.3. Gaussian Distribution

Very often an observed phenomenon in nature is the result of the sum of a large number of small interactions (molecular agitation causing thermal noise for example) and so the probability distribution of the observed event is often the result of a very large number of convolutions. It can be shown[1] that by performing enough multiple convolutions with almost any shaped distribution the final distribution tends to a definite shape, this being the Gaussian distribution.

$$p(x) = \frac{1}{\sqrt{(2\pi)}V} e^{-x^2/2V^2} \tag{1.8}$$

Where V is the r.m.s. value of the waveform, (sometimes termed the standard deviation, or V^2 is sometimes termed the variance).

Thus thermal noise is one of the many random variables in nature which take on this amplitude probability density function.

1.3.4. Averages of Random Variables

The average of any waveform may be obtained by taking any amplitude of the waveform and weighting it by the probability that this waveform is at that amplitude and summing all such possible weighted amplitudes. The average is denoted by placing a bar over the quantity, thus,

$$\bar{x} = \int_{-\infty}^{\infty} x\,p(x)\,\mathrm{d}x \tag{1.9}$$

This is also called the expectancy of x, $E(x)$

$$E(x) = \bar{x} = \int_{-\infty}^{\infty} x\,p(x)\,\mathrm{d}x \tag{1.10}$$

In the same way, the mean square value (or variance) of the variable may be written as

$$\overline{x^2} = E(x^2) = \int_{-\infty}^{\infty} x^2 p(x)\,\mathrm{d}x \tag{1.11}$$

A common alternative notation for $E(x^2)$ is μ_2. Similarly the nth moment of the probability distribution of the random variable x is written as,

$$\mu_n = \overline{x^n} = E(x^n) = \int_{-\infty}^{\infty} x^n p(x)\,\mathrm{d}x \tag{1.12}$$

If the waveform under consideration can be completely defined (i.e. is not a random variable) by the function $f(t)$, then of course its average may be written as,

$$\bar{x} = \lim_{T\to\infty} \frac{1}{2T} \int_{-T}^{+T} f(t)\,\mathrm{d}t \tag{1.13}$$

and,

$$\overline{x^n} = \lim_{T\to\infty} \frac{1}{2T} \int_{-T}^{+T} [f(t)]^n\,\mathrm{d}t \tag{1.14}$$

1.3.5. The Correlation Function

The correlation function is an extremely important function in communication engineering and is formed by taking the average of all the products of two points on a waveform separated by a parameter τ (usually time). Select a point of amplitude x_t on the waveform, the probability that a point τ seconds afterwards is $x_{t+\tau}$ is given by $p(x_t$ at time t, $x_{t+\tau}$ at time $t+\tau)$ and this may be written as $p(x_t, x_{t+\tau})$. Thus for one particular value x_t the average $x_{t+\tau}$ is found by integrating over all $x_{t+\tau}$ and the final average is obtained by integrating over all x_t. The final correlation function $R(\tau)$ is a function of τ, the separation between the points on the waveform. Thus,

$$R(\tau) = E(x_t . x_{t+\tau}) = \int_{-\infty}^{\infty} \int_{-\infty}^{\infty} x_t \, x_{t+\tau} p(x_t, x_{t+\tau}) \, dx_t \, dx_{t+\tau} \quad (1.15)$$

Again, if the waveform can be completely defined by the function $f(t)$, the correlation function may be written as

$$R(\tau) = \lim_{T \to \infty} \frac{1}{2T} \int_{-T}^{+T} f(t) f(t+\tau) \, dt \quad (1.16)$$

The function of equations 1.15 and 1.16 are also known as autocorrelation functions and autocovariance functions and in more rigorous treatments the distinction between these functions is important. Within the scope of this book the functions will be lumped together and termed the correlation function.

As the name implies the correlation function expresses the correlation between one part of a waveform and another or the correlation between one waveform and another.

It should therefore be noted that the expectation or average of the product of two variables can be represented in two ways depending on the nature of the variables. First

$$\mu_{12} = E(x_1 . x_2)$$

Here it is inferred that x_1 and x_2 are separate random variables and the expectation or average μ_{12} gives the correlation between the two, and is a constant.

Secondly

$$R(\tau) = E[x_t . x_{t+\tau}]$$

Here it is inferred that the correlation is between a point in time on a single waveform and another point τ seconds later on the same waveform, averaged over all points. The waveform may be either a random variable or a definable function and $R(\tau)$ is a function of τ. For instance, if τ is large then the variation of a point is likely to be almost independent of the variation of a point τ seconds away (depending on the type of function of course) and so the average of the product will be nearly zero. If $\tau = 0$ then the two points are coincident and the average yields the mean square value. In general therefore correlation functions have a maximum at zero and die away as τ increases[2]. For periodic functions, the correlation function is also periodic.

Example 1.2

Derive an expression for the correlation between $A \cos(\omega t+\phi)$ and $A \cos[\omega(t+\tau)+\phi]$

Solution

The expression to be evaluated corresponds to equation 1.16 and the correlation function is given by

$$R(\tau) = \lim_{T\to\infty} \frac{1}{2T} \int_{-T}^{+T} A^2 \cos(\omega t+\phi) \cos[\omega(t+\tau)+\phi]\, dt$$

$$= \lim_{T\to\infty} \frac{A^2}{2T} \int_{-T}^{+T} \frac{1}{2} [\cos(2\omega t+\omega\tau+2\phi)+\cos \omega\tau]\, dt$$

$$= \lim_{T\to\infty} \frac{A^2}{4T} \left[\frac{\sin(2\omega t+\omega\tau+2\phi)}{\omega} + t \cos \omega\tau \right]_{-T}^{T}$$

$$= \frac{A^2}{2} \cos \omega\tau$$

Note that the correlation function is independent of the phase ϕ and that the mean square value of the waveform (or the power in the waveform) is given by $R(0)$ i.e. $A^2/2$.

1.3.6. The Fourier Transform

Any waveform which varies with time can be expressed as the sum of a number of sine waves of differing frequency and phase with respect to each other. Thus, if there is a function in time $f(t)$ then it is usually possible to construct a function $F(f)$ which describes the waveform just as completely but this time in terms of its frequency components.

For example, periodic functions in time may be expanded into a series of sine waves by the Fourier series technique, and as a square wave consists of a fundamental and an infinite number of odd harmonics (three times the fundamental; five, seven times, etc.) when a square wave is passed through a low pass filter which only allows the fundamental to pass, the fundamental will be observed at the output; similarly any other of the harmonics would be observed by using suitable narrow band filters. Since such concepts as filtering, bandwidth and spectral distribution of energy are vital to communications, the ability to transform from axes of amplitude versus time (time domain) to axes of amplitude, amplitude density or power density versus frequency (frequency domain) is equally vital. The Fourier transform is the mathematical means by which such a transformation can take place. The transform used in this book is

$$F(f) = \int_{-\infty}^{\infty} f(t)\, e^{-j\omega t}\, dt \tag{1.17}$$

$$f(t) = \int_{-\infty}^{\infty} F(f)\, e^{j\omega t}\, df \tag{1.18}$$

where $\omega = 2\pi f$

However, with the random variables with which this book is concerned $f(t)$ cannot be expressed explicitly and therefore $F(f)$ cannot be found. The correlation function can be expressed explicitly however and it can be shown[2, 3] that the Fourier transform of this gives the spectral power density of the random variable.

$$S(f) = \int_{-\infty}^{\infty} R(\tau)\, e^{-j\omega\tau}\, d\tau \tag{1.19}$$

$$R(\tau) = \int_{-\infty}^{\infty} S(f)\, e^{j\omega\tau}\, df \tag{1.20}$$

where $\omega = 2\pi f$

$S(f)$ is the power density spectrum of the random variable and has the units—power per Hz. It is mathematically convenient to represent the spectrum in the positive frequency plane and the negative frequency plane as shown in Figure 1.5a. This arises from the concept of a vector of fixed direction varying sinusoidally in

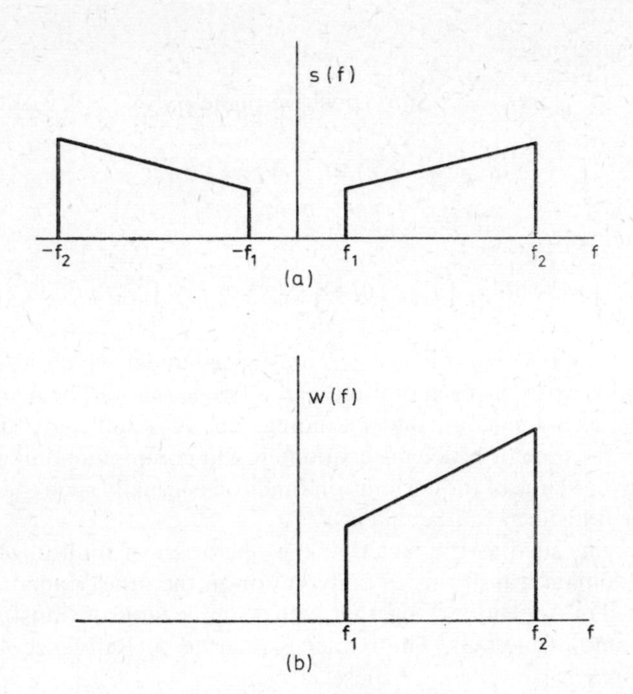

Figure 1.5 Power density spectra. (a) Double sided; (b) single sided

amplitude being represented by the sum of two vectors rotating at equal speeds one in the clockwise direction and the other in the anticlockwise direction.

The total power in the frequency band f_1 to f_2 shown in Figure 1.5a is given by

$$\int_{f_1}^{f_2} S(f)\,\mathrm{d}f + \int_{-f_2}^{-f_1} S(f)\,\mathrm{d}f = 2\int_{f_1}^{f_2} S(f)\,\mathrm{d}f \qquad (1.21)$$

since the contribution from the negative half plane must be added to that in the positive half plane.

The total power in the waveform is given by

$$\int_{-\infty}^{\infty} S(f)\,\mathrm{d}f \qquad (1.22)$$

It should be noted that the power density spectrum or power spectrum is sometimes represented in the positive plane only (Figure 1.5b) and in order that the total power should remain correct the height of the spectrum is doubled. Thus if $w(f)$ is the single sided power spectrum,

$$w(f) = 2\,S(f)\ \text{(positive plane only)} \tag{1.23}$$

$$\int_0^{\infty} w(f)\,df = \int_{-\infty}^{\infty} S(f)\,df = \text{Total Power} \tag{1.24}$$

and

$$\int_{f_1}^{f_2} w(f)\,df = \int_{-f_2}^{-f_1} S(f)\,df + \int_{f_1}^{f_2} S(f)\,df = 2\int_{f_1}^{f_2} S(f)\,df \tag{1.25}$$

Obviously the rigour of the circumstances under which Fourier transforms exist has been omitted and reference should be made to suitable texts for a full understanding, but it is sufficient to say that for the type of problems encountered in communication engineering the shape of the correlation function is usually such that the Fourier transform can be taken.

Theorems such as the fact that the operation of multiplication in one domain transforms to convolution in the other domain are tabulated in Appendix 1, together with a simple table of transforms for common functions. The reader is referred to Reference 4 for a comprehensive version of these.

1.3.7. The Characteristic Function

The fact that the probability that the sum of two independent variables with probability density functions $p_1(x)$ and $p_2(x)$ lies in a small element at z has been given previously by equation 1.7

$$p_{1+2}(z) = \int_{-\infty}^{\infty} p_1(\alpha)p_2(z-\alpha)\,d\alpha$$

which is recognised as the convolution integral. In the last section benefit was obtained from taking the Fourier transform of a close relation of the convolution integral, namely the correlation function (equation 1.16). This indicates that a useful parameter might be the Fourier transform of the probability density function. The Fourier transform of the probability density function is called the *characteristic function* of the parameter and can be written as,

$$M(jv) = \int_{-\infty}^{\infty} p(x)\,e^{jvx}\,dx \tag{1.26}$$

where v is the transform dummy variable corresponding to $-\omega$ of equation 1.17. Equation 1.26 can also be recognised under another name as the expectancy of e^{jvx}, cf. equation 1.10 thus

$$E(e^{jvx}) = \int_{-\infty}^{\infty} e^{jvx} p(x)\, dx \tag{1.27}$$

The joint probability density function for two variables x_1 and x_2 has been given as $p(x_1, x_2)$ and so the characteristic function may be written as,

$$\int_{-\infty}^{\infty}\int_{-\infty}^{\infty} p(x_1, x_2) e^{jv_1x} e^{jv_2x}\, dx_1\, dx_2 = E[e^{j(v_1x_1+v_2x_2)}] \tag{1.28}$$

The multivariate characteristic function (the characteristic function for many variables) for N variables $x_1, x_2, x_3, \ldots, x_N$ is

$$E\left[e^{j\sum_{n=1}^{N} v_n x_n}\right] \tag{1.29}$$

1.3.8. Properties of Gaussian Variables

In Section 1.3.3. it was stated that a variable x having the probability distribution

$$p(x) = \frac{1}{\sqrt{(2\pi)}} e^{-x^2/2} \tag{1.30}$$

is a Gaussian random variable, and thus it is defined.

The corresponding characteristic function [which is the Fourier transform of $p(x)$] also defines the variable and this is given by

$$M(jv) = e^{-v^2/2} \tag{1.31}$$

The joint probability distribution for two statistically independent Gaussian variables x_1 and x_2 is the product of their separate probabilities,

$$p(x_1, x_2) = \frac{1}{2\pi} \exp\left(-\frac{x_1^2}{2} - \frac{x_2^2}{2}\right) \tag{1.32}$$

If the Gaussian variables x_1 and x_2 are not statistically independent their joint probability is given by

$$p(x_1, x_2) = \frac{1}{2\pi(\mu_{20}\mu_{02} - \mu_{11}^2)^{1/2}} \exp\left[\frac{-\mu_{02}x_1^2 + 2\mu_{11}x_1x_2 - \mu_{20}x_2^2}{2(\mu_{20}\mu_{02} - \mu_{11}^2)}\right] \tag{1.33}$$

$$\mu_{20} = E(x_1^2) \quad \mu_{02} = E(x_2^2) \quad \mu_{11} = E(x_1 . x_2) \tag{1.34}$$

(see Reference 3, Section 8.2).

The corresponding joint characteristic function for Gaussian variables is

$$M(jv_1, jv_2) = \exp\left[-1/2(\mu_{20}v_1^2+2\mu_{11}v_1v_2+\mu_{02}v_2^2)\right] \quad (1.35)$$

In subsequent chapters use will be made of the multivariate probability distribution and the multivariate characteristic function for the Gaussian variables, $x_1, x_2, x_3, \ldots, x_N$.

The former is the probability of finding a Gaussian variable at a value x_1 another at a value x_2, etc. This probability distribution is given by

$$p(x_1, x_2, \ldots, x_N) = \frac{\exp\left[-\dfrac{1}{2|\mu|}\sum_{n=1}^{N}\sum_{m=1}^{N}|\mu|_{nm}x_nx_m\right]}{(2\pi)^{N/2}|\mu|^{1/2}} \quad (1.36)$$

where $|\mu|_{nm}$ is the cofactor of the element μ_{nm} in the determinant $|\mu|$ of the correlation matrix

$$[\mu] = \begin{bmatrix} \mu_{11} & \mu_{12} & \cdots & \mu_{1N} \\ \mu_{21} & & & \\ \vdots & & & \\ \mu_{N1} & \mu_{N2} & \cdots & \mu_{NN} \end{bmatrix}$$

this may be abbreviated to

$$p(x_1, x_2, \ldots, x_N) = [(2\pi)^N|\mu|]^{-1/2}\exp\{-\tfrac{1}{2}[\tilde{x}][\mu]^{-1}[x]\} \quad (1.37)$$

where $[x]$ is a column matrix and $[\tilde{x}]$ is its transpose.

Again these equations may be used as a definition of Gaussian variables.

The multivariate Gaussian characteristic function is given by

$$M(jv_1, jv_2, \ldots, jv_N) = \exp\left(-\frac{1}{2}\sum_{n=1}^{N}\sum_{m=1}^{N}\mu_{nm}v_nv_m\right) \quad (1.38)$$

This final generalised statement (which may also be regarded as a definition of Gaussian variables) enables a relationship between the expectancy of the product of a number of Gaussian variables $x_1, x_2, x_3, \ldots, x_N$ and the expectancy of pairs (i.e. correlation functions) to be established. This is very useful when a polynomial in $x(t)$ is to be multiplied by a polynomial in $x(t+\tau)$ as it allows the decomposition of multiple moments into manageable second moments (correlation functions).

The relationship is proved in Appendix 2 and is stated as

$$E(x_1.x_2.x_3 \ldots x_N) = 0 \quad \text{for} \quad N \quad \text{odd} \tag{1.39a}$$

$$= \sum_{\text{all pairs}} \left[\prod_{\substack{j \neq k \\ k=1 \\ j=1}}^{N} E(x_j.x_k) \right] \quad \text{for } N \text{ even} \tag{1.39b}$$

The right hand side of this equation means that the $N/2$-fold product of $E(x_j.x_k)$ is formed using all possible different combinations of pairs of x_j and x_k and these combinations are added.

As an example, the expectancy of the product of four Gaussian variables is given by

$$\begin{aligned} E(x_1.x_2.x_3.x_4) = \; & E(x_1.x_2)E(x_3.x_4) \\ & + E(x_1.x_3)E(x_2.x_4) \\ & + E(x_1.x_4)E(x_2.x_3) \end{aligned}$$

Example 1.3

Express the correlation between $x^3(t)$ and $x^3(t+\tau)$ in terms of the correlation between $x(t)$ and $x(t+\tau)$ where $x(t)$ is a Gaussian random variable.

Solution

It is required to expand $E[x^3(t)x^3(t+\tau)]$ and for simplicity $x(t)$ will be written as x and $x(t+\tau)$ as x_τ

$$E(x^3.x_\tau^3) = E(x.x.x.x_\tau.x_\tau.x_\tau) \tag{1.40}$$

It is now required to form products of all possible different sets of pairs. For brevity the components of the right hand side of equation 1.40 will be represented as a, b, c, etc., as shown below,

$$\begin{aligned} & E(a\,b\,c\,d\,e\,f) \\ & = E(ab)\,E(cd)\,E(ef) + E(ac)\,E(bd)\,E(ef) + E(bc)\,E(ad)\,E(ef) \\ & \quad + E(ab)\,E(ce)\,E(df) + E(ab)\,E(cf)\,E(de) + E(ac)\,E(be)\,E(df) \\ & \quad + E(ac)\,E(bf)\,E(de) + E(bc)\,E(ae)\,E(df) + E(bc)\,E(af)\,E(de) \\ & \quad + E(ad)\,E(be)\,E(cf) + E(ae)\,E(bf)\,E(cd) \\ & \quad + E(af)\,E(be)\,E(cd) + E(bd)\,E(ce)\,E(af) \\ & \quad + E(cd)\,E(be)\,E(af) + E(cd)\,E(ae)\,E(bf) \end{aligned}$$

Referring to equation 1.40 for the meanings of a, b, c, etc., we have $a = b = c = x$ and $d = e = f = x_\tau$. Thus,

$$E(x^3 . x_\tau^3) = 9E(x^2)\,E(x_\tau^2)\,E(x . x_\tau) + 6E^3(x . x_\tau) \tag{1.41}$$

1.4. USEFUL FUNCTIONS

1.4.1. The Impulse or Delta Function

The impulse or delta function results from taking a rectangular function as shown in Figure 1.6a which is of unit area and allowing the width to tend to zero while keeping the area constant. The height of the pulse tends to infinity as the width shrinks towards zero.

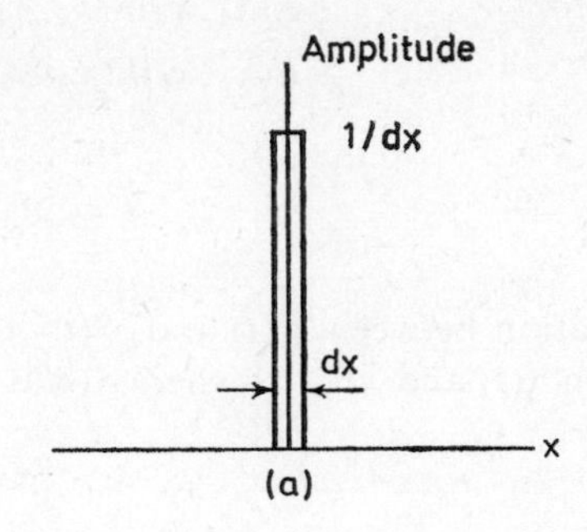

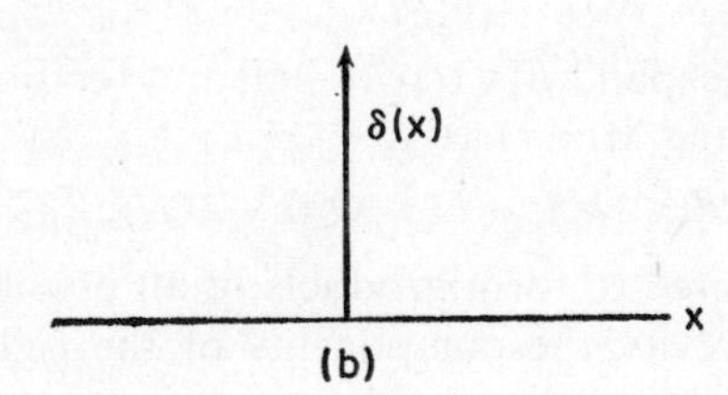

Figure 1.6 The delta function

The function at the position $x = 0$ is written as $\delta(x)$ and is represented by the vertical arrow shown in Figure 1.6b. Since the function has unit area,

$$\int_{-\infty}^{\infty} \delta(x)\,\mathrm{d}x = 1 \tag{1.42}$$

As an example of the occurrence of delta functions consider the power spectrum of the function

$$f(t) = A \sin \omega t$$

the total power or mean square value is $A^2/2$ and obviously all of the power in the spectrum is at frequency $f = \omega/2\pi$. The double sided power density spectrum must therefore consist of two delta functions one at $-f$ and one at $+f$ such that the total power obtained by integrating the spectrum as in equation 1.22 equals $A^2/2$. The magnitude of the delta function is therefore $A^2/4$ and the spectrum is shown in Figure 1.7. The same arguments could be used for cos ωt or any other phased sinusoid and in fact it can be

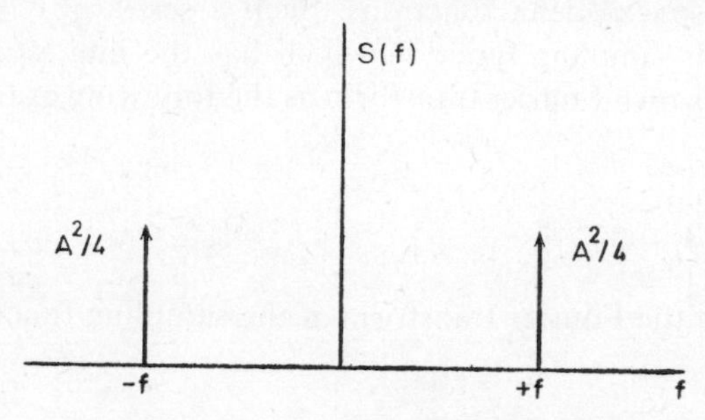

Figure 1.7 The power spectrum of a sinusoid

shown (see example 1.2) that the correlation function of any sinusoid of frequency ω is $A^2 \cos \omega t/2$. The power density spectrum is the Fourier transform of the correlation function and the dictionary of Appendix 1 shows that this is the two symmetrically displaced delta functions of Figure 1.7.

Note that $\delta(x)$ represents a delta 'spike' at $x = 0$ and so $\delta(x-a)$ represents a delta 'spike' at $(x-a) = 0$, i.e. at $x = a$. As before

$$\int_{-\infty}^{\infty} \delta(x-a)\, \mathrm{d}(x-a) = 1$$

The function $\delta(x/a-1)$ also represents a delta 'spike' at $x/a-1=0$ i.e. at $x = a$ but this time

$$\int_{-\infty}^{\infty} \delta(x/a-1)\, \mathrm{d}(x/a-1) = 1$$

or

$$\int_{-\infty}^{\infty} \delta(x/a-1)\, \mathrm{d}x = a$$

Thus

$$\delta(x/a-1) = a\, \delta(x-a) \tag{1.43}$$

Notice for example that since

$$\delta(\omega/2\pi-1) = 2\pi\delta(\omega-2\pi) \text{ (where } \omega = 2\pi f)$$

is a delta function at $\omega = 2\pi$ (or $f = 1$)

$$\delta(f) = 2\pi\delta(\omega)$$

1.4.2. The Sampling Function

In certain communication processes (and in particular the digital systems of Chapter 6) it is necessary to sample a waveform at regular intervals in order to communicate discrete information rather than continuous information. Mathematically this is equivalent to multiplying a waveform to be sampled by a series of regularly spaced delta functions. Such a series of delta pulses is termed the sampling function which has the interesting property of being its own Fourier transform as the following example shows.

Example 1.4

Show that the Fourier transform of the sampling function is itself.

Solution

Consider the pulse waveform shown in Figure 1.8a. Its Fourier transform is given by

$$A(f) = \frac{\sin \pi\tau f}{\pi\tau f} \frac{A\tau}{T} \sum_{n=-\infty}^{\infty} \delta(f-n/T)$$

Amplitude per Hertz.

This may be obtained by using the Fourier series technique to obtain the amplitude expression before the summation sign. The summation of delta pulses then converts this to an amplitude density spectrum.

If A is put equal numerically to $1/\tau$ then as τ contracts towards zero, the waveform tends to a series of delta pulses. In the frequency domain, the central height of the envelope of the delta pulses of Figure 1.8b remains constant but the width of the envelope increases.

In the limit the envelope becomes

$$\left.\frac{\sin \pi f\tau}{\pi\tau f}\right|_{\tau\to 0} = 1$$

Thus in the limit the transform may be written as

$$\sum \delta(t-nT) \text{ transforms to } \sum 1/T\, \delta(f-n/T)$$

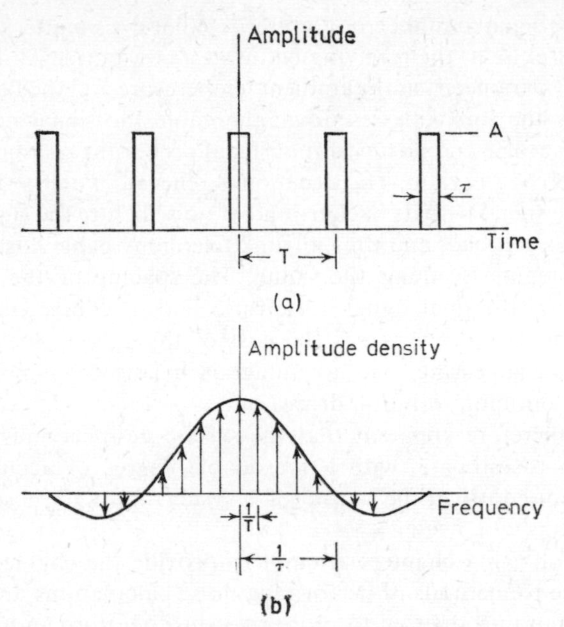

Figure 1.8 Amplitude density spectrum of a pulse train. (a) Time domain; (b) frequency domain

Each side of this transform may be rearranged according to equation 1.43 as follows

$$1/T \sum \delta(t/T-n) \text{ transforms to } \sum \delta(fT-n)$$

Theorem 3 of Appendix 1 (Similarity Theorem) now shows that the transform may be re-written as

$$\sum \delta(t-n) \text{ transforms to } \sum \delta(f-n)$$

So the sampling function $\sum \delta(t-n)$ is its own Fourier transform.

1.5. NOISE IN COMMUNICATION SYSTEMS

The basic mathematical tools for noise analysis have been given and in the subsequent chapters they will be used to show how systems are engineered. The whole problem of communication engineering centres around noise and obtaining adequate signal/noise ratios.

For example, a trunk telephony transmission along cable is transmitted with a certain power, and as the signal passes down the cable it becomes attenuated. This fact by itself does not matter since the signal could be re-amplified. What does matter is that the thermal noise at the receiving end is at a constant level depending on such parameters as the ambient temperature and the bandwidth. Thus as the signal passes down the cable the signal/noise ratio becomes worse and no amount of signal processing or amplification will improve the ratio. The situation is relieved of course by amplifying the signal with its background of noise before the signal/noise ratio gets too bad, and thus a trunk telephony cable has repeaters spaced regularly along the route. The spacing of the repeaters determines the final signal/noise ratio and of course the spacing of the repeaters determines the cost of the system and so, as is usual in engineering, one advantage is balanced against another and the optimum solution drawn out.

It is therefore apparent that the system designer must be in a position to estimate, with a reasonable degree of accuracy, the signal/noise ratio at the output of a system for a given set of parameters.

The following chapters attempt to provide the engineer with a reference to methods of performing noise calculations and also to provide an introduction to more rigorous literature and literature which extends the subject further.

Problems

1.1 Find expressions for, and sketch the probability density distributions for the following variables:

(a) A random variable equally likely to be between the values -10 and $+10$ V.

(b) A random variable taking values only between -8 and $+8$ V, whose probability density between -5 and $+5$ V is a constant which is twice the constant probability density elsewhere.

(c) A sawtooth waveform of peak-to-peak amplitude 4 V centred on 0 V.

(d) A sinusoid of peak value 6 V centred on 0 V.

(e) A square wave between the values -5 and $+5$ V.

1.2 (a) What is the probability that the random variable of Problem 1.1a is greater than 6 V? (Ans. 1/5)

(b) What is the probability that the random variable of Problem 1.1b lies between $+3$ and $+7$ V? (Ans. 3/13)

(c) What is the probability that the sinusoid of Problem 1.1d is greater than 5·9 V? (Ans. 0·055)

1.3 (a) Find from first principles the mean value of a random variable which is equally likely to be between the values 0 and 10 V. (Ans. 5 V)

(b) Find the mean square value of a random variable whose probability of being at a certain value is proportional to the voltage and exists between 0 and 10 V. (Ans. 50 V)

(c) Show that for statistically independent variables the mean of their sum is the sum of their means.

1.4 Use the dictionary of Fourier transforms and the Fourier transform theorems of Appendix 1.1 to prove the following transforms.

$$\cos \omega_c t \longrightarrow 1/2\,\delta(f-f_c)+1/2\,\delta(f+f_c)$$

$$e^{j\omega_c t} \longrightarrow 2\pi\,\delta(\omega-\omega_c)$$

An isosceles triangle of unit height and unit area $\longrightarrow \dfrac{\sin^2 \pi f}{(\pi f)^2}$

$$\left.\begin{matrix} t & |t|<1/2 \\ 0 & |t|>1/2 \end{matrix}\right\} \longrightarrow \frac{j}{2\pi}\frac{\mathrm{d}}{\mathrm{d}f}\left(\frac{\sin \pi f}{\pi f}\right)$$

$$\left.\begin{matrix} -1 & t<0 \\ 1 & t>0 \end{matrix}\right\} \longrightarrow \frac{j}{\pi f}$$

1.5 Show that if $x(t)$ is a Gaussian random variable whose autocorrelation function is $R(\tau)$ then the autocorrelation function of the variable $[x(t)]^2$ is

$$[R(0)]^2+2[R(\tau)]^2$$

A Gaussian random variable of unit total power and having a flat power density spectrum extending from $-3{\cdot}84$ to $+3{\cdot}84$ MHz (and zero elsewhere) is passed through a perfect square law device. Calculate the power spectrum of the variable at the output.

(Ans. Delta function of 1 W at $f = 0$ and a ramp from 7·68 MHz to 0·26 W/MHz)

References

1. MIDDLETON, D., *Introduction to Statistical Communication Theory*, (Section 7.7), McGraw-Hill, New York (1960)
2. PROF. DR. ING. HABIL and LANGE, F. H., *Correlation Techniques*, (Section 2.2), Iliffe, London (1967)
3. DAVENPORT, W. B. and ROOT, W. L., *Random Signals and Noise*, McGraw-Hill, New York (1958)
4. BRACEWELL, R. M., *The Fourier Transform and its Applications*, McGraw-Hill, New York (1965)

Chapter 2

THERMAL NOISE

2.1. INTRODUCTION

At normal temperatures the free electrons within a conductor are in a state of agitation, the amount of agitation depending upon the absolute temperature. As each electron moves it gives rise to a very small current or voltage at the terminals of the conductor and the summation of the random movements of the large number of similar electrons gives rise to a current or voltage waveform known as thermal noise.

The waveform of thermal noise is, therefore, made up of small contributions from a large number of sources and it has been indicated in Chapter 1 (Sections 1.3.2 and 1.3.3) that the probability distribution for the amplitude of the waveform is therefore Gaussian. Thus the probability of finding the amplitude of the waveform within the gap, width dx, centred on the amplitude x is given by

$$p(x)\,\mathrm{d}x = \frac{1}{\sqrt{(2\pi)}\,V}\,\mathrm{e}^{\frac{-x^2}{2V^2}}\,\mathrm{d}x \tag{2.1}$$

where V is the r.m.s. value of the waveform.

In communication theory it is of interest to know how the noise power density from a conductor or resistor varies with frequency i.e. to know the shape of the noise power spectrum. In principal this could be determined by connecting the resistor (at room temperature say) in turn to a number of ideal noise free filters of equal bandwidths and centred on a range of different frequencies. It would be found that the power from each filter is the same thus indicating that the power density spectrum is the same value for all frequencies. The problem may be tackled theoretically, for example, by kinetic methods where the electrons are considered to move in the fashion of a random walk around the atomic lattice or by thermodynamical methods[1]. In both cases it is found that for most practical frequencies (i.e. frequencies less than about 10^{12} Hz) the power density spectrum is flat. The magnitude of the spectrum may be expressed in terms of the maximum power available (P_{max}) in a bandwidth B, this being obtained by considering the power to be dissipated in a load which is matched to the noise source.

$$P_{max} = kTB \tag{2.2}$$

where k is Boltzmann's constant ($1{\cdot}37 \times 10^{-23}$ J/K)
T is the absolute temperature (K)
B is the bandwidth under consideration (Hz).

The r.m.s. voltage developed across a resistor R ohms is therefore given by

$$V_{\text{r.m.s.}} = \sqrt{(4kTBR)}\ \text{V} \tag{2.3}$$

and the noise source may be represented by the equivalent circuits of Figure 2.1.

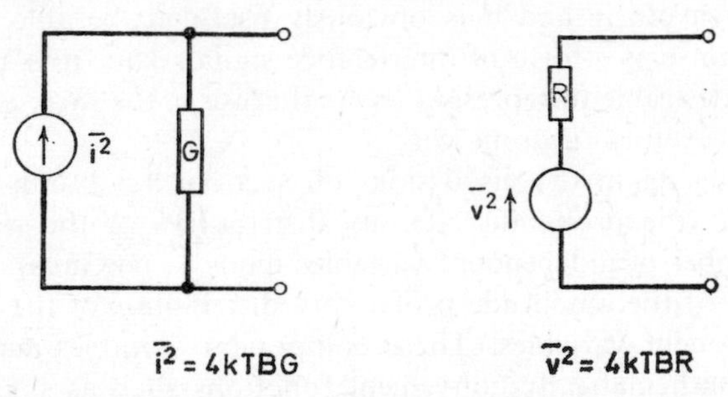

Figure 2.1 Equivalent circuits for noise in resistors

The double sided noise power density spectrum $N_0(f)$ is given from equation 2.2.

$$N_0(f) = kT/2 \tag{2.4}$$

and it should be noted that this spectrum and the maximum available power itself are independent of the resistance of the conductor. Thus the available noise power from any object is given by kTB.

Any object which has varying currents in it acts as an aerial and radiates power and the received noise power in an aerial matched to its receiver is also kTB, where T refers this time to the average temperature of the bodies that the aerial is 'looking' at.

Line-of-sight radio link aerials generally point along the surface of the earth and 'see' mainly the warm earth, and so T is of the order of 300 K. Communication satellite ground station aerials on the other hand point out into space where the average temperature is very much lower and so T in this case is very low. For example, for a paraboloid aerial pointing vertically through a dry atmosphere, $T = 2{\cdot}4$ K.

2.2. METHODS OF REPRESENTING THERMAL NOISE

Since the exact waveform of thermal noise cannot be predicted it is impossible to write down a function for it. However, as the discussions of the previous section and of Chapter 1 (Sections 1.3.5 and 1.3.6) show, it is possible to define the power spectrum of thermal noise, and hence its autocorrelation function can also be defined. These definitions are used in the analytical calculation techniques shown in this book, but in some calculation techniques indicated in Chapters 3 and 5 (Monte-Carlo computer techniques for example) the actual waveform of the thermal noise is represented. Also the development of the interference formulae of Chapter 3 uses a representation of the interfering signal in the form of a random waveform and it is obviously useful to be able to treat thermal noise as a form of interference signal. Thus it is apparent that it is desirable to represent thermal noise in the form of a time domain waveform in some way.

Recalling again the discussions of Sections 1.3.2 and 1.3.3 of Chapter 1, the probability density distribution of the sum of a large number of independent variables tends to be Gaussian independently of the amplitude probability distribution of the individual component variables. These component variables may therefore be mathematically convenient functions such as sine waves. Thus, provided the component sine waves are uncorrelated and a sufficiently large number are taken, the sum of the sine waves will be a random-like waveform with a Gaussian amplitude probability density distribution.

The power density spectrum of the representation of the noise depends on how the frequencies of the component sine waves are chosen. For example, a band of noise of flat power spectrum between the frequencies f_0 and f_m may be represented by the sum of a large number of sine waves of equal amplitude and of frequencies incrementally increasing from f_0 to f_m, each assigned an initial random phase angle

i.e.

$$\text{Representative Waveform} = \sum_{n=1}^{n=N} \sin(\omega_n t + \phi_n) \tag{2.5}$$

where $\omega_1 = 2\pi f_o$

$\omega_N = 2\pi f_m$

ϕ_n is a random phase angle

and where N is a very large number, the magnitude of which determines the closeness of the representation of the waveform to the noise waveform.

Another representation of the same band of noise can be given by assigning a different set of initial random phase angles.

Frequently in the analysis of communication systems it is only necessary to consider noise over a relatively narrow bandwidth (over the pass-bandwidth of a radio receiver for example) and in this case it is sometimes convenient to consider the component variables making up the representative model for noise to be sine waves of the same frequency. In order for the arguments concerning multiple convolutions to apply (Chapter 1, Section 1.3.2) the component sine waves must be uncorrelated. This condition was met in the previous model by making each of the components a slightly different frequency and in this model the requirement can be met by giving each component sine wave a continously varying random phase angle. If the frequency of all the component sine waves is ω which corresponds to the centre frequency of the narrow band of noise, then the sum of the sine waves may be expressed as

$$\text{Representative Waveform} = R(t) \sin [\omega t + \phi(t)] \qquad (2.6)$$

$R(t)$ is a random variable of magnitude greater than zero, corresponding to how the particular component random phase angles add at any instant. $\phi(t)$ is a varying random phase angle taking any value between 0 and 2π radians. Since this representative waveform is made up of the sum of many independent components its amplitude probability distribution will be Gaussian. The probability distribution of $R(t)$ is not Gaussian but Rayleigh as demonstrated in Example 2.1.

If a narrow band of noise centred on a high frequency is observed on an oscilloscope, the waveform resembles a sine wave with random amplitude modulation and random phase modulation. This, of course, corresponds to the form of equation 2.6 and explains why this representative waveform is particularly useful.

Example 2.1

Demonstrate that a narrow band of Gaussian noise centred on the frequency ω may be represented by the waveform $x_c(t) \cos \omega t + x_s(t) \sin \omega t$ where x_c and x_s are independent Gaussian variables and hence show that the probability distribution of $R(t)$ in equation 2.6 is Rayleigh.

Solution

Equation 2.6 shows that a narrow band of Gaussian noise centred on the frequency ω may be represented as

$$\text{Representative Waveform} = R(t)\,[\sin \omega t+\phi(t)] \qquad 2.7$$

Figure 2.2 shows a phasor diagram of the addition of the component sine waves all at the same frequency but with random phase angles. The phasor diagram is imagined to be rotating at ω rad/s.

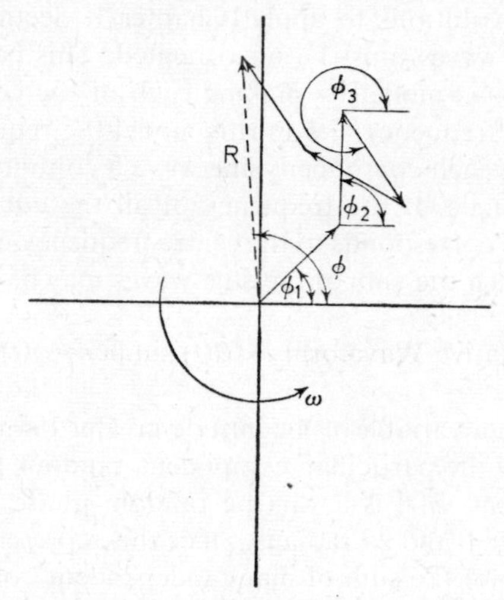

Figure 2.2 Representation of thermal noise by the sum of randomly phased sinusoids

Equation 2.7 may be written as

$$R(t) \sin [\omega t+\phi(t)] = R(t) \sin \phi(t) \cos \omega t+R(t) \cos \phi(t) \sin \omega t \qquad (2.8)$$

From Figure 2.2 it can be seen that

$$R(t) \sin \phi(t) = \sum_{n=1}^{N} [\sin \phi_1(t)+\sin \phi_2(t)+\ldots \sin \phi_n(t)+\ldots] \qquad (2.9)$$

$$R(t) \cos \phi(t) = \sum_{n=1}^{N} [\cos \phi_1(t)+\cos \phi_2(t) \ldots \cos \phi_n(t)+\ldots] \qquad (2.10)$$

where $\phi_1(t)$, $\phi_2(t)$ etc. are the random phase angles of the component sine waves making up the representative waveform of equation 2.7. The right hand sides of equations 2.9 and 2.10 show that $R(t)\cos\phi(t)$ and $R(t)\sin\phi(t)$ are made up of a large number of independent components and hence the probability distribution of the sum in each case is Gaussian.

Writing $x_c(t) = R(t)\sin\phi(t)$ and $x_s(t) = R(t)\cos\phi(t)$ it is seen from equation 2.8 that an alternative representative waveform is given by

$$\text{Representative Waveform} = x_c(t)\cos\omega t + x_s(t)\sin\omega t \quad (2.11)$$

where $x_c(t)$ and $x_s(t)$ are Gaussian variables. Since

$$E\,[\sin\phi(t).\cos\phi(t)] = 0 \qquad \text{(see Appendix 4)}$$

then $R(t)\sin\phi(t)$ and $R(t)\cos\phi(t)$ are uncorrelated and $x_c(t)$ and $x_s(t)$ are therefore statistically independent. The probability of finding $x_c(t)$ in a gap dx_c centred on x_c and the probability of finding x_s in a gap dx_s centred on x_s is given by

$$p(x_c, x_s)\,dx_c.dx_s = \frac{1}{2\pi V^2}\exp\left(\frac{-x_c^2 - x_s^2}{2V^2}\right) dx_c.dx_s \quad (2.12)$$

Since $x_c^2 + x_s^2 = R^2$ (where R is a value of the waveform $R(t)$) and since $dx_c.dx_s$ in Cartesian co-ordinates corresponds to $R.dR.d\phi$ in polar co-ordinates, the probability of finding the waveform in a magnitude gap dR centred on a magnitude R and phase angle gap $d\phi$ centred on ϕ is given by

$$p(R, \phi)\,dR.d\phi = \frac{R}{2\pi V^2}\exp\left(-\frac{R^2}{2V^2}\right) dR.d\phi \qquad (2.13)$$

Since the angle ϕ only appears as a differential and therefore its probability distribution is a constant it follows that the angle ϕ is equally likely to take any possible angle between 0 and 2π and its probability distribution must be

$$p(\phi)\,d\phi = \frac{d\phi}{2\pi} \qquad (2.14)$$

Since also ϕ and R are independent

$$p(R, \phi) = p(R)p(\phi) = \frac{R}{V^2}\exp\left(-\frac{R^2}{2V^2}\right)\frac{1}{2\pi}$$

and so

$$P(R) = \frac{R}{V^2}\exp\left(-\frac{R^2}{2V^2}\right) \qquad (2.15)$$

This is called a Rayleigh Distribution.

2.3. Signal/Noise Ratio

The parameter of fundamental importance in calculating the quality of a communication system is the output signal/noise ratio. Noise may be regarded as negative information or a destroyer of information and so the signal/noise ratio gives a measure of the amount of useful information contained in the output. The signal/noise ratio is usually defined in terms of the signal power and the noise power and is given by

$$\mathrm{S/N} = \frac{\text{Signal Power at Terminals}}{\text{Noise Power at Terminals}}$$

it is usually quoted in decibels (dB)

$$\mathrm{S/N} = 10\log_{10}\frac{\text{Signal Power at Terminals}}{\text{Noise Power at Terminals}} \tag{2.16}$$

Much of the analysis of this book concerns signals in the form of many channels in frequency-division-multiplex and the signal and noise at the output of a device is expressed in terms of the signal power density spectrum at the output and the noise power density spectrum at the output. These spectra may have different amplitudes at different frequencies as shown in Figure 2.3 and therefore

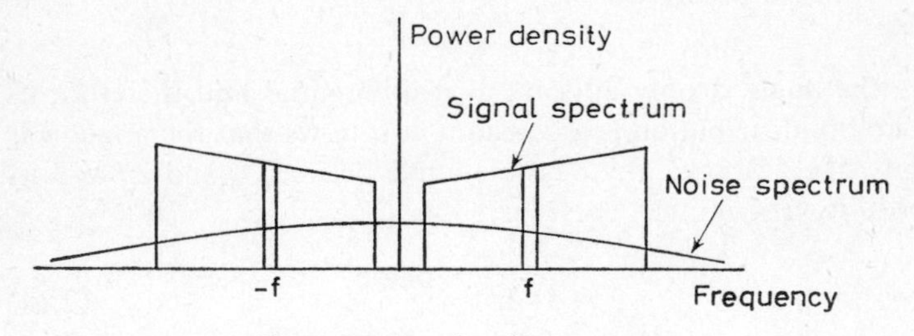

Figure 2.3 Signal and noise spectra

the S/N ratio at one frequency within the signal spectrum may be different from another. Therefore at a particular frequency,

$$\mathrm{S/N} = \frac{S(f)}{N_0(f)} = \frac{\begin{array}{c}\text{Signal Power in an infinitely narrow bandwidth}\\ \text{centred on the frequency } f\end{array}}{\text{Noise Power in the same bandwidth}} \tag{2.17}$$

For any general bandwidth f_1 to f_2

$$S/N = \frac{\int_{f_1}^{f_2} S(f)\,df}{\int_{f_1}^{f_2} N_0(f)\,df} = \frac{\text{Signal Power in the bandwidth } f_1 \text{ to } f_2}{\text{Noise Power in the bandwidth } f_1 \text{ to } f_2}$$

2.4. NOISE FACTOR AND NOISE TEMPERATURE

2.4.1. Noise Factor

Any device which transmits or processes a signal introduces additional noise. This may be shot noise in vacuum tube devices, thermal effects in transistor amplifiers and thermal noise in any lossy device. Thus, as a signal passes through a communication system the signal/noise ratio is degraded and the amount of degradation introduced by a device is measured by its Noise Factor (F). This may be defined as

$$F = \frac{\text{Signal/noise ratio at input of device}}{\text{Signal/noise ratio at output of device}} \tag{2.18}$$

Figure 2.4 shows an amplifier of gain G and the noise generated within the amplifier is represented by the noise N_A at the input. The noise factor is thus

$$F = \frac{S_{in}/N_{in}}{S_{out}/N_{out}} = \frac{S_{in}/N_{in}}{GS_{in}/G(N_{in}+N_A)}$$

$$F = \frac{N_{in}+N_A}{N_{in}} \tag{2.19}$$

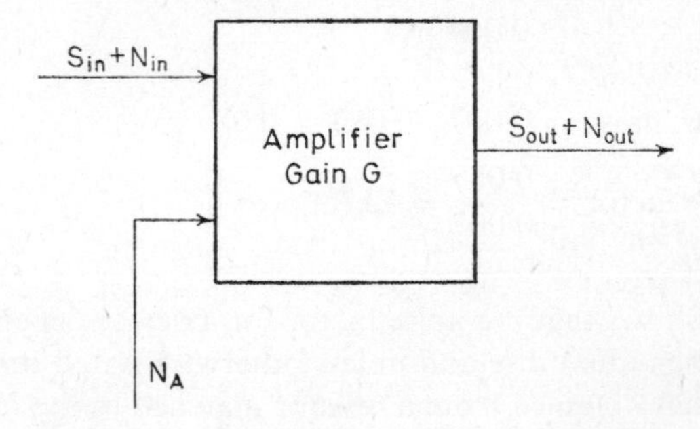

Figure 2.4 Representation of noise in an amplifier

Thus the noise factor may be expressed as

$$F = \frac{\text{Actual noise power at the output of the device}}{\text{Noise power if the device is assumed to be noiseless}}.$$

Example 2.2

The noise generated within an amplifier of bandwidth 5 MHz may be represented by a noise power at the input of the amplifier of 0·082 pW. Calculate the noise factor of the amplifier if it is fed from (a) a matched signal source at a temperature 300 K, (b) a matched signal source cooled to 100 K.

Solution

(a) The power available from the matched source at room temperature is kTB.

$$\text{Therefore input noise} = 1{\cdot}37 \times 10^{-23} \times 300 \times 5 \times 10^{6}\ \text{W}$$

$$= 0{\cdot}021\ \text{pW}$$

Effective input noise $= 0{\cdot}021 + 0{\cdot}082 = 0{\cdot}103$ pW from equation 2.19

$$\text{Noise factor} = \frac{0{\cdot}103}{0{\cdot}021}$$

$$= 4{\cdot}9$$

(b) In this case the input noise is given by

$$\text{Input noise} = 1{\cdot}37 \times 10^{-23} \times 100 \times 5 \times 10^{6}$$

$$= 0{\cdot}007\ \text{pW}$$

Effective input noise $= 0{\cdot}007 + 0{\cdot}082 = 0{\cdot}089$ pW

$$\text{Noise factor} = \frac{0{\cdot}089}{0{\cdot}007} = 13\ \text{(approx)}$$

Example 2.2 shows that the noise factor for a device depends upon the level of the input noise and unless otherwise stated this level is taken to be that obtained from a resistor matched to the input and at room temperature (290 K).

2.4.2. Noise Temperature

An alternative method of defining the degradation of signal/noise ratio in a device, which overcomes this difficulty, uses a parameter known as the noise temperature. This is the temperature by which a matched resistor at the input of the device must be raised to account for the noise within the device.

If the input noise is $N_{in} = kT_{in}B$, the additional noise contributed by the device can be represented by $N_A = kT_AB$, where T_A is the noise temperature of the device expressed in Kelvin.

Total noise at output of device of given gain (G) is $Gk(T_{in}+T_A)B$ whereas the noise at output of device, if assumed noiseless is $GkT_{in}B$.

Therefore: $$F = \frac{T_{in}+T_A}{T_{in}} = 1+\frac{T_A}{T_{in}}$$

If T_{in} is taken as 290 K

$$F = 1+\frac{T_A}{290} \tag{2.20}$$

Example 2.3

Calculate the noise factor for a parametric amplifier whose noise temperature is (a) 8 K, (b) 10 K

Solution

From equation 2.20

(a) $$F = 1+\frac{8}{290} = 1{\cdot}0276$$

(b) $$F = 1+\frac{10}{290} = 1{\cdot}0345$$

Example 2.3 serves to show that the noise factor is not a convenient way of describing the noise performance of very low noise temperature devices since the results are all approximately 1.0. For this reason the concept of noise temperature is nearly always used to describe low noise devices.

2.4.3. Cascading of Stages

Figure 2.5 shows two amplifiers of gain G_1 and G_2 and noise factors F_1 and F_2 connected in series. The noise generated internally in each amplifier is N_{A1} and N_{A2} respectively and N_s is the standard input noise power from which the noise factors are defined.

For amplifier 1 $\quad F_1 = \frac{N_{A1}+N_s}{N_s} = 1+\frac{N_{A1}}{N_s}$

For amplifier 2 $\quad F_2 = \frac{N_{A2}+N_s}{N_s} = 1+\frac{N_{A2}}{N_s}$

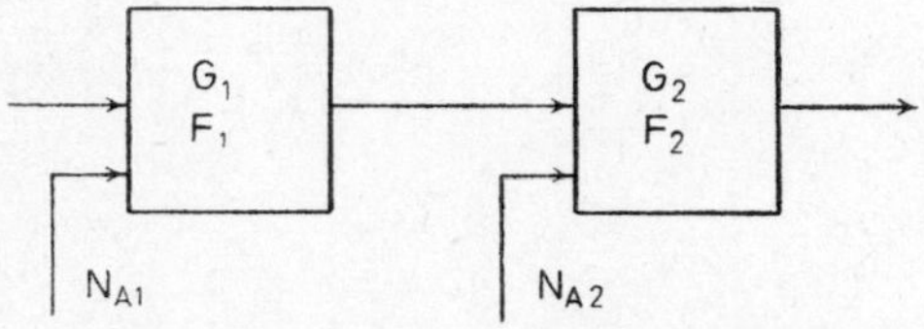

Figure 2.5 Noise in cascaded amplifiers

To find the overall noise factor for the two devices it is necessary to refer the noise introduced by the second amplifier to the input of the first and then assume that the second amplifier is noiseless.

The additional noise added to the input of the first amplifier is N_{A2}/G_1

i.e. $$\text{Additional noise} = \frac{N_{A2}}{G_1} = \frac{(F_2-1)N_s}{G_1}$$

The overall noise figure is now given by

$$F_{1+2} = 1+\left(N_{A1}+\frac{N_{A2}}{G_1}\right)\frac{1}{N_s}$$

i.e. $$F_{1+2} = F_1+\frac{F_2-1}{G_1} \tag{2.21}$$

In general for many amplifiers,

$$F = F_1+\frac{F_2-1}{G_1}+\frac{F_3-1}{G_1G_2}+\frac{F_4-1}{G_1G_2G_3}+\dots \tag{2.22}$$

Example 2.4

A communication satellite system ground station receiver consists of an aerial, waveguide run from the aerial to the input of the first amplifier, low noise first amplifier, and succeeding stages. Derive an expression for the system noise temperature in terms of the following parameters.

T_{system} = System noise temperature referred to the input terminals of the first low noise amplifier.
T_{sky} = Noise temperature of the sky at which the aerial is pointing.
T_{feed} = Physical temperature of the waveguide feeder.
L = Loss introduced by the waveguide run expressed as the ratio of signal power in/signal power out.
T_1 = Noise temperature of the first amplifier (low noise).
G_1 = Gain of the first amplifier (low noise).
F = Noise factor of the succeeding stages.

Solution

(a) Sky

The noise temperature of the sky is T_{sky} and therefore the noise received by the aerial will be $KT_{sky}B$. If the feeder is assumed to be noiseless, then the sky noise appearing at the input terminals of the first amplifier will be $(kT_{sky}B)/L$.
Therefore,

$$\text{Sky noise temperature referred to input terminals of first amplifier} = \frac{T_{sky}}{L}$$

(b) Feeder (see also Problem 2.2)

The waveguide feeder (or any other type of feeder) may be considered as an ideal loss free noiseless transmission line with some of the signal power tapped off at the input to correspond to the loss. Figure 2.6 shows such a system where the power fed in is unity, the power at the output is $1/L$ and the quantity of power $1-[1/L]$

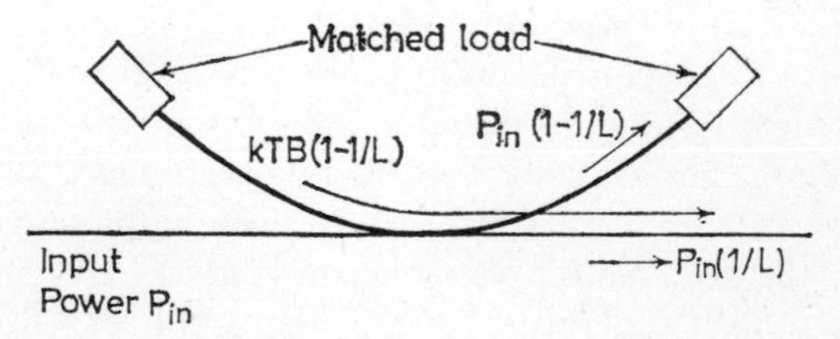

Figure 2.6 Representation of an attenuator by an ideal waveguide coupler

is coupled out and absorbed in a matched load. The matched load absorbing the portion $1-[1/L]$ however will generate noise power corresponding to a noise temperature T_{feed} which is the physical temperature of the waveguide. Thus by symmetry a noise power $kT_{feed}B[1-1/L]$ will be coupled into the output of the feeder. Therefore,

$$\text{feeder noise temperature referred to input terminals of first amplifier} = T_{feed}\left(1-\frac{1}{L}\right)$$

(c) Succeeding Stages

The noise temperature of the succeeding stages is derived from Equation 2.20

$$F = 1+\frac{T_2}{290}$$

where T_2 is the noise temperature of the succeeding stages

$$T_2 = (F-1)\,290$$

This noise temperature must be referred to the input of the first amplifier by dividing by the gain of the first amplifier.

Therefore,

$$\text{Noise temperature of succeeding amplifiers referred to the input of the first amplifier} = \frac{(F-1)\,290}{G_1}$$

The total system noise temperature is now given by

$$T_{system} = \frac{T_{sky}}{L}+T_{feed}\left(1-\frac{1}{L}\right)+T_1+\frac{(F-1)290}{G_1}$$

Note the following points

(a) Since T_{feed} is large compared with T_{sky} (provided the aerial is pointing into space) the waveguide run should be designed to make L as close to unity as possible. For this reason the waveguide run between the aerial feeder and the first amplifier is kept as short as possible and frequently the waveguides employ wave modes other than the dominant mode (over-moded).

(b) The noise temperature of the first amplifier (T_1) is critical since this is a direct contribution to the system temperature. For

this reason the first amplifier of a low noise system such as a communication satellite system is usually either a cooled maser or cooled parametric amplifier and the noise temperatures are typically only a few Kelvin.

(c) The noise factor of the succeeding stages is not critical provided the gain of the first amplifier is reasonably large (this may be 1 000 or more).

(d) The quality of communication receiver systems is often expressed in terms of the ratio of the gain of the system aerial to the system noise temperature referred to the input of the first amplifier.

$$\text{Quality of receive system} = \frac{\text{Gain of aerial}}{T_{\text{system}}}$$

(e) Because of the definition of noise factor, the result for the noise temperature of a feeder cannot be substituted directly into equation 2.20 to find the noise factor of a feeder or attenuator. Where the term noise factor is used it is usual for the source to be taken at the same physical standard temperature (T_s) as the attenuator. The signal/noise ratio at the input in this case will be S_{in}/kT_sB. The signal is attenuated by a factor L while the noise at the output remains the same and so the signal/noise ratio at the output is S_{in}/LkT_sB. The noise factor is therefore given by

$$F = L \text{ (for the above assumption).}$$

2.5. A SUMMARY OF THE EFFECTS OF THERMAL NOISE ON VARIOUS TYPES OF RECEIVER

2.5.1. Analogue Receivers

In order to perform system calculations, it is necessary to know how the thermal noise at the input terminals of the receiver demodulator is translated to noise in the final output. Chapter 3 deals with the problem of interference of a general nature into various analogue systems. The results can be used for the particular case of thermal noise which can be considered as form of interference.

In Chapter 3, the performance of a receiver is described in terms of a transfer factor X such that,

$$X = \frac{S_{out}/N_{out}}{S_{in}/N_{in}}$$

In Table 2.1 the results of Chapter 3 are taken with the following assumptions.

Table 2.1

Demodulator type	$\frac{S_{out}/N_{out}}{S_{in}/N_{in}} = X$	*Section of book covering relevant theory*
Amplitude Modulation. Single sideband, suppressed carrier, demodulated by a synchronous detector	1	Equation 3.24
Amplitude Modulation. Double sideband, 100% modulation with ratio of peak-to-r.m.s. value of modulating signal h/v	$\frac{2}{[1+(h/v)^2]}$	Equation 3.26 and Example 3.3
Frequency Modulation. f_d = r.m.s. frequency deviation of the carrier by a single 1 mW test tone. α = frequency of channel under consideration in the modulating signal. B = bandwidth of receiver. f_B = bandwidth of channel. $pe(\alpha)$ = pre-emphasis characteristic. S_{out}/N_{out} = ratio of 1 mW test tone to thermal noise level in a channel.	$\left(\frac{f_d}{\alpha}\right)^2 \left(\frac{B}{f_B}\right) pe(\alpha)$	Example 3.5

(a) The information being transmitted by the communication system has a flat power spectrum extending from zero frequency to a frequency f_m.

(b) The interference power spectrum is flat and extends over the input bandwidth of the receiver.

Example 2.5

Ten sections of ideal transmission line each of loss 80 dB are connected by repeaters each of gain 80 dB and of noise factor 3 dB. The end section of the line is connected to a receiver of noise factor 3 dB and this is required to receive telephone channels in frequency-division-multiplex at a signal/noise ratio of 35 dB.

There is a choice of modulation method between single sideband-suppressed carrier amplitude modulation (s.s.b.-s.c.) and frequency modulation (f.m.)

Calculate:

(a) The transmitter power required to transmit the maximum possible number of telephone channels if the bandwidth of the transmission line is restricted to 400 kHz.

(b) If the same transmitter power is used as in case *(a)* calculate the number of telephone channels that can be transmitted if the transmission bandwidth is unrestricted.

Use the following parameters for frequency modulation.
Minimum allowable signal/noise at input of the receiver = 13 dB, consisting of a threshold signal/noise at input of the receiver of 10 dB and a threshold margin of 3 dB.
Pre-emphasis allowance for top baseband channel 4 dB.

Solution

A diagram of the communication system is given in Figure 2.7 and it is first necessary to obtain the overall noise factor. Considering the first length of line of gain $1/L$ (times) and the repeater at the

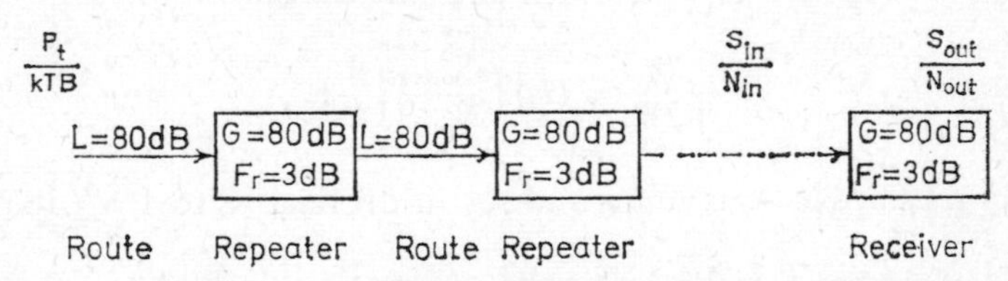

Figure 2.7 Communication system for Example 2.5

end of this section of gain L (times) and noise factor F_r, the two items in cascade give an overall noise factor (from equation 2.22 and note (*e*) page 39) of

$$F = L + L(F_r - 1)$$
$$= LF_r$$

Subsequent sections of line with their associated repeaters and the last section of line with the receiver also of noise factor F_r give an overall noise figure of

$$LF_r + (LF_r - 1) + (LF_r - 1) \ldots$$

Thus for N sections of line, the total noise factor (Taking $LF_r \gg 1$) is

$$F = NLF_r$$

i.e. $$\text{Overall noise factor} = 10+80+3 \text{ dB}$$
$$= 93 \text{ dB}$$

(a) In this part there is no limit to the power available but there is a limit to the bandwidth (400 kHz) and so the modulation method s.s.b.-s.c. which requires the minimum bandwidth is chosen, see Section 3.4.1.

Signal/noise ratios are defined at three points in Figure 2.7 (1) the ratio in the received telephone channel (S_{out}/N_{out}), (2) the ratio of signal power/noise power at the input terminals of the demodulator (S_{in}/N_{in}), and (3) the overall transmitter power/noise power ratio at the input to the first line section (P_t/kTB).

The first two ratios are related by the interference transfer factor X.

$$S_{out}/N_{out} = X[S_{in}/N_{in}]$$

The second two ratios are related by the noise factor for the system between the input and the demodulator

i.e. $$\left(\frac{P_t}{kTB}\right)\left(\frac{N_{in}}{S_{in}}\right) = F$$

(Here B is the 400 kHz bandwidth of the transmission).

Thus $$\frac{P_t}{kTB} = (F)\left(\frac{S_{out}}{N_{out}}\right)\left(\frac{1}{X}\right)$$

For s.s.b.-s.c. X is unity and so, in dB relative to 1 W (dBW),

$$\begin{array}{l} P_t+204-56 = 93 \quad +35 \quad -0 \\ \quad\quad (kT) \quad (B) \quad (F) \quad (S_{out}/N_{out}) \, (X) \end{array}$$

The transmitter power required is given by

$$P_t = -20 \text{ dBW}$$

or $$\text{Transmitter power} = 10 \text{ mW}$$

Allowing 4 kHz for each telephone channel then the 400 kHz of bandwidth allows transmission of 100 channels.

(b) In this part the bandwidth is now unrestricted and so frequency modulation which exchanges bandwidth for power is chosen, see Section 3.5.

As the input signal/noise ratio to a frequency demodulator decreases, the output signal/noise ratio decreases proportionally according to the factor X until a certain threshold input signal/noise is reached. Below this threshold the output signal/noise ratio worsens considerably more than that predicted by the factor X. This threshold input signal/noise ratio is usually taken as 10dB and a margin of 3dB is added as a safety precaution. Thus the minimum value that S_{in}/N_{in} can take is 13 dB.

i.e.
$$\frac{S_{in}}{N_{in}} = 13 \text{ dB}$$

Thus
$$\frac{P_t}{kTB} = \underset{(F)}{13+93}\text{dB}$$

For a transmitter power of 10 mW

$$\underset{(P_t)\ (kT)}{-20+204} \quad \underset{\text{(bandwidth)}}{-10 \log B} = 106$$

Which gives
$$B = 63 \text{ MHz}$$

If the ratio S_{out}/N_{out} is required to be 35 dB then the required value of X is 22 dB.

From the table of Section 2.5.1.

$$X = (f_d/\alpha)^2 \, (B/f_B) \, pe(\alpha)$$

X is worst in the top baseband channel and so α is put equal to f_m, the top baseband frequency.

Thus
$$158{\cdot}5 = (f_d/f_m)^2 \frac{63 \times 2{\cdot}52}{0{\cdot}004}$$

where all the frequencies are expressed in MHz.

i.e.
$$f_d/f_m = 0{\cdot}063 \tag{2.23}$$

Another relationship between f_d and f_m is required in order to obtain f_m. This may be obtained by writing the r.f. bandwidth of the transmission in terms of the peak frequency deviation caused by the complete baseband and the modulating frequency. Thus if ΔF is the r.m.s. frequency deviation caused by all of the channels, a peak to mean ratio (a) may be used to obtain the peak deviation. Work on peak clipping in Chapter 6 shows that a reasonable figure for (a) might be 3.

Thus
$$B = 2(3\Delta F + f_m)$$

This is an approximate formula for the bandwidth if for no other reason than the vagueness of the bandwidth of an f.m. signal.

If f_d is the r.m.s. deviation caused by one speech channel, then the r.m.s. deviation caused by the whole baseband is given by

$$\Delta F^2 = N(f_d)^2 \quad \text{where } N \text{ is the number of channels.}$$

Since $$N = f_m/0{\cdot}004$$

$$B = 2\left[3\left(\frac{f_m}{0{\cdot}004}\right)^{1/2} f_d + f_m\right]$$

i.e. $$31{\cdot}5 = 47{\cdot}4(f_m)^{1/2} f_d + f_m \tag{2.24}$$

From equations 2.23 and 2.24 it is found that f_m is approximately 5.6 MHz, corresponding to about 1400 telephone channels.

2.5.2. Digital Receivers

The response of digital receivers to thermal noise is complicated and cannot be represented by a simple transfer factor like that used for analogue receivers. There is also additional noise in the output of a digital system due to quantising noise, and it should be noted that this plays an important part in the system design. The methods of calculating the noise due to quantising and thermal effects are given in detail in Chapter 6 and an example of a system calculation is given as example 6.7. For the sake of completeness, however, a summary of the effect of thermal noise only on the more common types of digital systems follows.

Noise at the input to a demodulator gives rise to the possibility of the demodulator making an error in interpreting the digit being sent. For example in a binary system where there are two signal output possibilities (analogous to 'dot' and 'dash' in Morse code) the noise may cause the receiver to demodulate the input as one signal when it should have been the other. The degree to which this happens may be expressed in terms of the probability that an error will be produced (P_e). For given noise conditions, the probability of error depends on the way in which the digits are transmitted from the transmitter to the receiver i.e. the type of secondary modulation used. Table 2.2 gives expressions for the probability of error for common types of secondary modulation.

The amount of noise appearing in the analogue output of the decoder section of a receiver, for a given probability of error, depends upon the method of coding used (the primary digital modulation method). Table 2.3 shows how the decoder output signal/noise is related to the probability of error in the digits for the two types of primary digital modulation discussed in Chapter 6.

Table 2.2

Secondary modulation (Binary systems only)	*Formula for the probability of error* (P_e)	*Section of book covering relevant theory*
Frequency shift keying signals $\sin \omega_1 t$ and $\sin \omega_2 t$ *Amplitude shift keying* signals $\sin \omega t$ and 0 or a d.c. level and 0	$P_e = \frac{1}{2}[1 - \text{erf}\,(E/4N_0)^{1/2}]$	Example 6.3
Phase shift keying signals $\sin(\omega t+\phi)$ and $\sin(\omega t+\phi+\pi/2)$	$P_e = \frac{1}{2}[1 - \text{erf}\,(E/2N_0)^{1/2}]$	Problem 6.4

where

P_e = Probability of error

E = Signal energy i.e. signal power × duration of signal bit

N_0 = Two-sided noise power spectral density

erf (x) = The error function given by

$$\text{erf}\,(x) = \frac{1}{\sqrt{(2\pi)}} \int_{-\sqrt{2x}}^{\sqrt{2x}} e^{-y^2/2}\,dy$$

Tables of this function are readily available and it is plotted in Figure 6.21. and 5.1c

Table 2.3

Primary Modulation	*Noise due to errors / Signal power* $\left(\frac{N_e}{S}\right)$	*Section of book covering relevant theory*
Pulse code modulation (No companding)	$\frac{4\,P_e a^2}{3}$	Section 6.4.4
Pulse code modulation (With companding)	Error noise given in graphical form	Figure 6.14
Sigma-delta modulation	$\frac{8P_e a^2 f_m}{f_c}$	Section 6.5.3.

where

$$a = \text{The ratio } \frac{\text{clipping voltage of analogue signal}}{\text{r. m.s. value of analogue signal}}$$

f_m = Maximum analogue modulating frequency

f_c = System clock frequency

Problems

2.1 Use the characteristic function for a Gaussian variable (equation 1.31) to show that the amplitude distribution of the sum of two thermal noise sources (equation 1.7) of mean square values V_1^2 and V_2^2 has a Gaussian distribution of mean square value $(V_1^2+V_2^2)$

2.2 The Reciprocity theorem states that the current produced in any branch of a linear network, by an e.m.f. in any other branch, equals the current in the other branch which would result if the e.m.f. was transferred to the first branch.

Use this theorem to show that the maximum available noise appearing across any two terminals, due to the resistive components in a network is $kTB\alpha$.

k is Boltzmann's constant.

T is the network temperature.

B is the bandwidth of the matched load.

α is the proportion of power absorbed in the network when a signal is applied at the same two terminals.

2.3 A baseband signal is clipped at 3 times its r.m.s. value and used to amplitude modulate a carrier (double sideband) with a maximum depth of modulation. The total signal power at the input to the demodulator is 1 mW and the total thermal noise power is 2 μW. Calculate the baseband signal to thermal noise ratio at the output of the demodulator.

(Ans. 20 dB)

2.4 300 channels of telephony are assembled in frequency division multiplex and used to frequency modulate a carrier. The r.f. bandwidth of the transmission is restricted to 16·8 MHz. Calculate the signal/noise ratio required at the input to the demodulator if the ratio of average signal power to noise in the top baseband channel is to be 35dB and there is a pre-emphasis advantage of 6 dB for the top baseband channel. Calculate also

the ratio of the carrier power to the system noise temperature required at the input to the demodulator.
(Note. f_d should be taken as the r.m.s. frequency deviation caused by the average signal power and not that caused by a test tone, and that f_d may be calculated according to equation 2.24.)

(Ans. $S_{in}/N_{in} = 11{\cdot}5$ dB, $S_{in}/T = -144{\cdot}8$ dB)

References

1. MIDDLETON, D., *Introduction to statistical communication theory*, McGraw-Hill, New York (1960)

Chapter 3

ANALOGUE MODULATION METHODS AND INTERFERENCE

3.1. THE INTERFERENCE PROBLEM

In most of the present communication systems the demand for independent channels is rising. For example, in broadcast systems the number of different programmes transmitted simultaneously is increasing both on national and international scales and the number of telephone calls being made is increasing in a similar way. In these and other systems, therefore, methods of multiplexing or sharing must be devised whereby independent communication channels can be provided. As sharing becomes more widespread the risk of noise due to interference between one transmission and another increases and design calculations are necessary to keep the noise to a tolerable level.

In general there are three methods of multiplexing involving the sharing or division of frequency, space and time. The division of the radio frequency spectrum into separate bands for various communication systems and the sub-division of these bands is perhaps the most obvious of the three. For instance, the various broadcast services use different carrier frequencies and one service is selected from another by altering the frequency of reception of the receiver. In some bands the radio frequency spectrum is almost completely used up and it is acceptable to have services using approximately the same carrier frequency provided there is sufficient distance separating them. Thus the propagation space is also divided up for multiplexing purposes. Time division multiplex is not so commonly used in communication systems mainly because of the equipment complexity needed. As digital systems become more commonplace so time division multiplex will find more use. There is no problem of interference between channels sharing time, although of course there is a limit on the number of channels which can share a given system (see Chapter 6).

The signal/interference noise level, at the output of a receiver of a particular communication system can be considered to depend upon one or more of three main factors:

1. the ratio of the wanted signal power to the ratio of the unwanted signal power at the input terminals of the receiver, S_{in}/N_{in}.

2. The information carried by the wanted and the interfering systems (e.g. telephony, television, data) and the modulation method used in each system.

3. The frequency separation between the wanted and the unwanted signals.

The first of these factors is determined by many parameters connected with the way in which the propagation space is divided. Obviously the level of the signals at the receiver terminals depends upon the power of the wanted transmitter and the power of the interfering transmitter. The gains of the transmitting and receiving aerials and their polar diagrams are other factors. The various parameters may be summarised by the equation:

$$S_{in}/N_{in} = P_{rw} - (P_{tu} + G_{tu} + G_{ru} - L) \text{ dB}$$

where

P_{rw} = Signal power at the receiver terminals (wanted).
P_{tu} = Power transmitted by the interfering transmitter (unwanted).
G_{tu} = Gain of the interfering transmitter aerial in the direction of the interference path (affecting unwanted signal).
G_{ru} = Gain of the receiving aerial in the direction of the interference path (affecting unwanted signal).
L = Propagation loss between isotropic aerials situated at the positions of the interfering transmitter and the receiver.

The second and third of the three main factors are best grouped together in a factor designated X such that

$$S_{out}/N_{out} = X + S_{in}/N_{in} \text{ dB}$$

where

S_{out}/N_{out} = Ratio of signal to interference noise power, dB.
X = Receiver interference transfer factor or interference reduction factor, dB.

X thus expresses the manner in which the receiver of the wanted signal rejects the unwanted signal. If for instance there is a large frequency difference between the carriers of the wanted and the unwanted systems and if the receiver has a relatively small bandwidth, then there will be good rejection of the interference and X will be large.

3.2. THE INTERFERENCE TRANSFER FACTOR

The interference transfer factor X expresses the ability of a receiver to reject interference. As already indicated this ability can arise, for instance, because the interfering signal falls near or outside the skirts of the receiver passband filter and in most cases the rejection caused by this mechanism is high and interference calculations are unnecessary. If the interfering signal falls within the receiver passband then other factors affect the performance of the receiver. These factors such as the information carried and the modulation methods used by both the wanted and the interfering signals are considered in the following analysis. The modulation methods considered are restricted to analogue modulation of sinusoidal carriers (see Section 1.2.2).

A radio frequency sinusoidal carrier may be analogue modulated in two basic ways (Section 1.2.3) the modulating signal may be used to alter the amplitude of the carrier (amplitude modulation) or the phase of the carrier (phase or frequency modulation) and so in general the modulated carrier may be expressed as

$$f(t)\cos\left(\omega_c t+\phi(t)\right)$$

where $f(t)$ or $\phi(t)$ may vary with time but not usually both.

Since the unwanted signal (designated by the suffix u) is merely added to the wanted signal (suffix w) at the receiver input terminals the input $s(t)$ to the receiver may be expressed as,

$$s(t) = f_w(t)\cos\left(\omega_c t+\phi_w(t)\right)+f_u(t)\cos\left(\omega_c t+\omega_D t+\phi_u(t)\right) \tag{3.1}$$

where ω_D is the frequency difference between the wanted and the unwanted carriers.

Depending on the design of the receiver, this composite wave will either be treated as a single amplitude modulated carrier or a single phase modulated carrier. It is desirable therefore to express equation 3.1 in the form,

$$s(t) = A(t)\cos\left(\omega_c t+\phi_w(t)+\mu(t)\right) \tag{3.2}$$

If the components of equation 3.1 are represented by vectors as in Figure 3.1 where $\beta = \left(\omega_c t+\phi_w(t)\right)$ and $\alpha = (\omega_c+\omega_D)t+\phi_u(t)$,

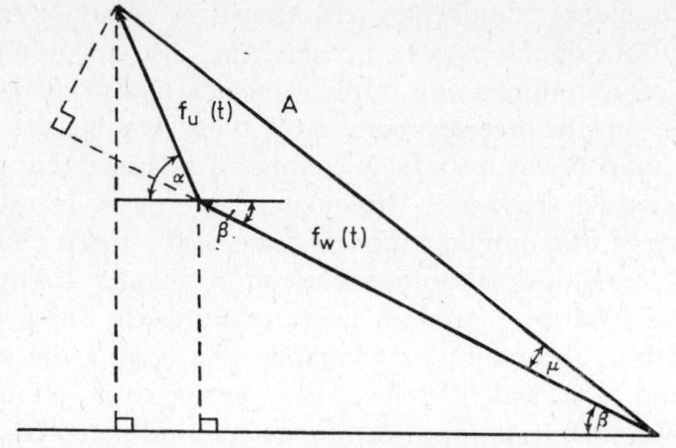

Figure 3.2 Vectorial representation of the interference problem

then it can be seen that

$$A^2 = [f_w(t)+f_u(t)\cos(\alpha-\beta)]^2 + [f_u(t)\sin(\alpha-\beta)]^2$$

i.e.

$$A^2 = f_w^2(t)+2f_w(t)f_u(t)\cos(\alpha-\beta)+f_u^2(t) \tag{3.3}$$

and

$$\mu = \tan^{-1}\frac{f_u(t)\sin(\alpha-\beta)}{f_w(t)+f_u(t)\cos(\alpha-\beta)} \tag{3.4}$$

The treatment of equations 3.3 and 3.4 obviously depends upon the type of modulation used in both the wanted and the unwanted systems. The following analysis classes the treatments of these equations according to the modulation method used by the wanted system and falls into three sections. Single sideband-suppressed carrier amplitude modulation, double sideband amplitude modulation, and frequency modulation. Single sideband systems are generally used for point to point communication using synchronous detection while double sideband modulation is generally used for broadcasting. Frequency modulation is used for broadcast and for point to point trunk communication (radio-relay systems).

The radio frequency spectrum of these systems plays an important part in the calculation of the interference transfer factor (see Chapter 2, Section 2.3) and before considering the solution of equations 3.3 and 3.4 methods of calculating the r.f. spectra are given.

3.3. THE BASEBAND SPECTRUM

The r.f. spectrum of a modulated carrier depends to a large extent on the nature of the modulating signal. This may be a single speech channel, music, or a television signal in the case of broadcast systems or blocks of multiplex telephony or television in trunk radio systems. Data signals from computers, etc., also account for a large amount of communication traffic. The very nature of information itself prevents the precise prediction of these signals and so approximations must be made and it is commonly assumed that the modulating signal is a sine wave. This model can be used to estimate the performance of communication systems of the single channel type for example. Blocks of multiplex telephony can be closely approximated by a band of random noise covering the same frequency bandwidth as the multiplex telephony and having the same total power and empirical rules have been evolved for estimating that total power[1]. Only the spectra of carriers modulated by noise-like waveforms will be considered here.

The modulating signal, termed the baseband, is represented by a band of noise of Gaussian amplitude probability distribution and extending from a minimum modulating frequency f_0 to a maximum modulating frequency f_m. The double sided power spectrum of the baseband is shown in Figure 3.2a. If the total mean square value of the modulating signal is V^2 then the power spectrum of the baseband $S_b(f)$ is given by:

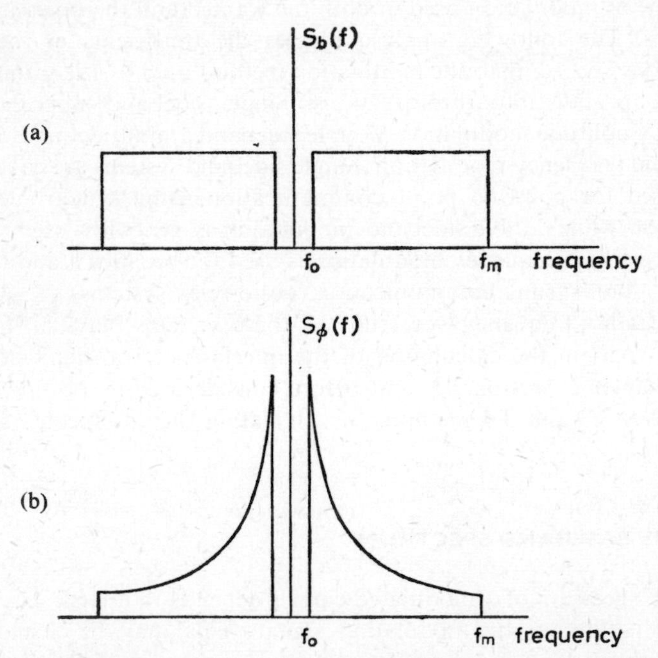

Figure 3.2 Baseband spectra. (a) The baseband spectrum; (b) the equivalent phase modulating spectrum

$$S_b(f) = \frac{V^2}{2(f_m - f_0)} \tag{3.5}$$

$$\text{for } -f_m < f < -f_0 \quad f_0 < f < f_m$$

$$S_b(f) = 0 \text{ elsewhere}$$

In the case of frequency modulation, a voltage in the baseband is converted to a change in frequency of the carrier, and so to avoid the use of conversion constants, instead of expressing the baseband power spectrum in terms of the mean square voltage, the mean square frequency deviation is used. This is written as ΔF^2 and so equation 3.5 becomes,

$$S_b(f) = \frac{\Delta F^2}{2(f_m - f_0)} \text{ (Hz)}^2 \text{ per Hz} \tag{3.6}$$

The basic equation to be considered in the case of frequency modulation is in the form,

$$s(t) = \cos\left[\omega_c t + \phi(t)\right]$$

and it can be seen that $\phi(t)$ is a noise signal phase modulating the carrier. Thus for frequency modulation the modulating signal $M(t)$ is $\mathrm{d}\phi(t)/\mathrm{d}t$ and,

$$\phi(t) = \int M(t)\,\mathrm{d}t$$

If equation 3.6 is integrated in the time domain an equivalent baseband will be obtained, which, when applied to a phase modulator gives the same output as the spectrum of equation 3.6 applied to a frequency modulator. This equivalent baseband spectrum is therefore given by

$$S_\phi(f) = \frac{\Delta F^2}{2(f_m - f_0)\omega^2} \quad \text{(cycles)}^2 \text{ per Hz}$$

i.e.

$$S_\phi(f) = \frac{\Delta F^2}{2(f_m - f_0) f^2} \quad \text{(radians)}^2 \text{ per Hz} \tag{3.7}$$

The shape of this spectrum is shown in Figure 3.2b and it should be noted that this baseband shape may be distorted by a pre-emphasis function $pe(f)$ which is included to improve the signal/noise ratio at the higher frequency end of the baseband.

In dealing with phase modulation an important quantity is the total mean square phase deviation caused by the modulating signal (P^2). This is obtained by integrating the power spectrum of the signal applied to a phase modulator over all frequencies. Thus,

$$P^2 = \int_{f_0}^{f_m} \frac{\Delta F^2}{(f_m - f_0) f^2}\,\mathrm{d}f = \frac{\Delta F^2}{f_m f_0} \quad \text{(radians)}^2 \tag{3.8}$$

(where pre-emphasis is used, this should be included in the integrand).

Equation 3.8 shows, for example, that a baseband to be frequency modulated with a given mean square frequency deviation ΔF^2 can have a very large value of mean square phase deviation P^2 or a small value of P^2 depending on the value of f_0. In fact it is quite reasonable for f_0 to extend down to zero frequency in which case for any value of ΔF^2, P^2 is infinite.

3.4. THE SPECTRA OF AMPLITUDE MODULATED SYSTEMS

3.4.1. Single Sideband Supressed Carrier. (s.s.b.-s.c.)

Single sideband modulation involves the translation of one sideband of the baseband to r.f. frequency and thus either the positive or the negative side of the baseband is multiplied by a carrier

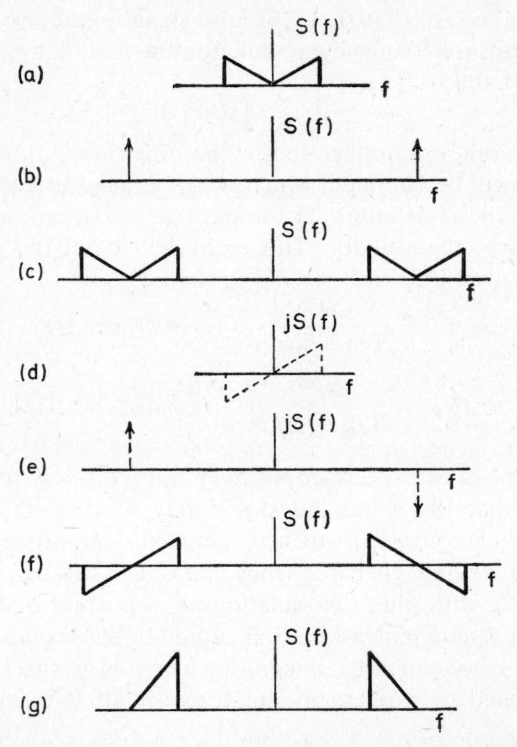

Figure 3.3 Formation of s.s.b.-s.c. signals. Spectrum of: (a) modulating signal $f(t)$*; (b)* $\cos \omega_c t$*; (c)* $f(t) \cos \omega_c t$*; (d)* $f(t)$ *with each frequency component shifted by* $(\pi/2) = f'(t)$*; (e)* $\sin \omega_c t$*; (f)* $f'(t) \sin \omega_c t$*; (g)* $f'(t) \sin \omega_c t + f(t) \cos \omega_c t$*.*

($\cos \omega_c t$). This can be achieved in practice by multiplying the complete signal by $\cos \omega_c t$ and then filtering out one of the sidebands or preferably by using a phase shift method. In the latter method the modulating signal is multiplied by $\cos \omega_c t$ shifting the spectrum in the usual way. The modulating signal also has each of its components phase shifted by $\pi/2$ and this has the effect of multiplying the spectrum in the positive frequency plane by $-$j and in the negative plane by $+$j, and this signal is then multiplied by $\sin \omega_c t$. Since the frequency spectrum of $\sin \omega_c t$ is $j\delta(f+f_c)-j\delta(f-f_c)$ addition of these two products produces a single sideband only as shown in Figure 3.3.

For the simple model to be used of a baseband with a flat spectrum, the spectrum of the single sideband system with a supressed carrier of frequency f_c is simply a rectangle extending from f_c+f_0 to f_c+f_m or from f_c-f_m to f_c-f_0.

3.4.2. Double Sideband Modulation

Double sideband modulation could be achieved by merely multiplying the double sided baseband by the carrier frequency $\cos \omega_c t$. Since there is no carrier present in this type of modulation, the carrier must be accurately regenerated in the receiver resulting in complicated receiver circuitry. The difficulty is overcome by inserting a large carrier component obtained simply by using the amplitude of the modulating signal to vary the amplitude of the carrier. Thus the modulated signal can be written as

$$s(t) = [1+f(t)] \cos \omega_c t \tag{3.9}$$

The level of the carrier relative to the modulating signal $f(t)$ must be such that its amplitude never falls below zero when amplitude modulated by the signal otherwise distortion will occur. Thus the minimum peak value of $1+f(t)$ must be greater than zero. If $f(t)$ is a signal with a Gaussian amplitude distribution this is clearly impossible and a compromise must be taken between the noise caused by clipping of the peaks when $[1+f(t)] < 0$ (or when the maximum amplitude of the carrier is reached) and the signal to thermal and/or interference noise which obviously worsens as the r.m.s. value of $f(t)$ becomes smaller. A graph of signal to clipping noise ratio is given in Chapter 6 (Figure 6.10) and this can be used to determine the optimum level of the modulating signal with respect to the carrier.

Again, calculation of the double sideband spectrum is straightforward for the flat baseband model used here and a diagram of the spectrum is given in Figure 3.4.

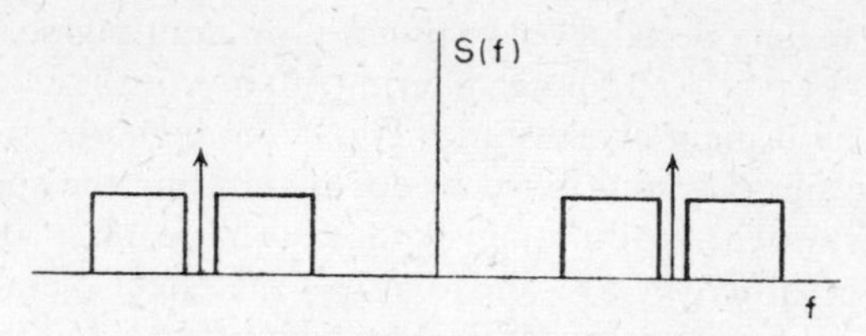

Figure 3.4 The spectrum of d.s.b. signals

3.5. THE SPECTRA OF FREQUENCY MODULATED SYSTEMS

3.5.1. General

The problem is to find the power spectrum of a function of the type,

$$s(t) = \cos[\omega t + \phi(t)]$$

where $\phi(t)$ is a noise signal representing the baseband.

It is shown in Appendix 3 that the spectrum of $s(t)$ about the carrier frequency $G_s(f)$ may be expressed as the sum of the spectra of $\cos\phi(t)$ and $\sin\phi(t)$ (denoted as $C_\phi(f)$ and $S_\phi(f)$ respectively in the Appendix)

i.e.
$$G_s(f) = C_\phi(f) + S_\phi(f) \tag{3.10}$$

A method of finding the power spectrum therefore reduces to finding the autocorrelation function of $\cos\phi(t)$ and $\sin\phi(t)$ and then taking the Fourier transform (see Chapter 1. Section 1.3.6 for an explanation of this technique). The autocorrelation function of $\cos\phi(t)$ for example is given by definition as,

$$R_c(\tau) = E[\cos\phi(t)\cos\phi(t+\tau)] \tag{3.11}$$

Equations 3.10 and 3.11 can be solved by considering the characteristic function of the bivariate Gaussian distribution which is defined in Chapter 1, Section 1.3.8.

$$\begin{aligned} M(jv_1, jv_2) &= E[e^{jv_1\phi(t)}e^{jv_2\phi(t+\tau)}] \\ &= \exp\left\{-\tfrac{1}{2}R_\phi(0)[v_1^2 + v_2^2 + 2\varrho(\tau)v_1v_2]\right\} \end{aligned} \tag{3.12}$$

where the correspondence between correlation functions and averages has been used as explained in Section 1.3.5. Note that $\varrho(\tau)$ is the normalised autocorrelation function $R_\phi(\tau)/R_\phi(0)$ and that in this context $R_\phi(0)$ is the 'power' of $\phi(t)$ i.e. the mean square phase deviation P^2. Putting $v_1 = 1$ and $v_2 = -1$

$$M(j, -j) = E\{[\cos\phi(t) + j\sin\phi(t)][\cos\phi(t+\tau) - j\sin\phi(t+\tau)]\}$$

and since products like $\sin \phi(t) \cos \phi(t+\tau)$ are uncorrelated,

$$M(j, -j) = E[\cos \phi(t) \cos \phi(t+\tau)] + E[\sin \phi(t) \sin \phi(t+\tau)]$$

which is the correlation function corresponding to $G_s(f)$, see equation 3.10. Thus from 3.12

$$\begin{aligned} M(j, -j) &= E[\cos \phi(t) \cos \phi(t+\tau)] + E[\sin \phi(t) \sin \phi(t+\tau)] \\ &= \exp\,[-R_\phi(0) + R_\phi(\tau)] \end{aligned}$$

Thus $R_s(\tau)$, the autocorrelation function of $s(t)$, is given by

$$R_s(\tau) = \exp\,[-R_\phi(0) + R_\phi(\tau)] \qquad (3.13)$$

and $G_s(f)$, the power spectrum of $s(t)$ about the carrier, is given by the Fourier transform of this expression

$$G_s(f) = \int_{-\infty}^{\infty} \exp\,[-R(0) + R(\tau)]\, e^{-j\omega\tau}\, d\tau \qquad (3.14)$$

Various methods have evolved for evaluating expressions like equations 3.13 and 3.14 and the most important of these follow together with other methods of obtaining spectra of frequency modulated systems. Considerable emphasis will be placed on the method of successive convolution, not because it is felt that this is particularly the best from a computational point of view but because it gives the best insight into how the frequency modulation spectrum is made up.

3.5.2. Method of Successive Convolution[2 4]

Equation 3.13 may be expanded as follows by putting $R_\phi(0) = P^2$

$$R_s(\tau) = e^{-P^2}\left[1 + \frac{R_\phi(\tau)}{1!} + \frac{R_\phi(\tau)^2}{2!} + \frac{R_\phi(\tau)^3}{3!} + \ldots\right] \qquad (3.15)$$

This equation may now be transformed term by term. The transform of the autocorrelation function of the baseband is the power spectrum of the baseband, and the powers of the autocorrelation function are transformed to the appropriate numbers of convolutions (see Appendix 1). Adopting the notation $\overset{n}{*}$ to mean the convolution process repeated $(n-1)$ times, the expression for the power spectrum of the modulated signal becomes,

$$G_s(f) = e^{-P^2}\left[\delta(f)+\frac{S_\phi(f)}{1!}+\frac{S_\phi(f)\overset{2}{*}S_\phi(f)}{2!}+\frac{S_\phi(f)\overset{3}{*}S_\phi(f)}{3!}+\ldots\right] \tag{3.16}$$

$S_\phi(f)$ is the baseband spectrum as applied to a phase modulator and if $w_\phi(f)$ is the normalised spectrum such that $S_\phi(f) = P^2 w_\phi(f)$ then the following result is obtained,

$$G_s(f) = e^{-P^2}\left[\delta(f)+P^2 w_\phi(f)+\frac{(P^2)^2}{2!}w_\phi(f)\overset{2}{*}w_\phi(f)\right.$$
$$\left.+\frac{(P^2)^3}{3!}w_\phi(f)\overset{3}{*}w_\phi(f)+\ldots\right] \tag{3.17}$$

and $$w_\phi(f)\overset{n}{*}w_\phi(f) = 1$$

The first term of this expression, $\delta(f)$, is a delta function and shows that there is a tone at the carrier frequency of power e^{-P^2}. Subsequent terms show that the spectrum is made up of the baseband (first order sidebands), the baseband spectrum convolved with itself (second order sidebands), and higher order sidebands of increasing orders of convolution. The total power in the nth order sidebands is given by $e^{-P^2}(P^2)^n/n!$ which is the nth term of the Poisson distribution. Thus if P^2 is small, most of the power of the spectrum is contained in the first few sidebands and the power in the tone at the carrier frequency is also large. If P^2 is large most of the spectral power is in the higher order sidebands.

In general, analytical solutions for the higher order sidebands are difficult but the expression 3.17 lends itself to simple numerical methods of evaluation. The baseband, with any form of pre-emphasis, may be represented by ordinates and then a successive convolution process used to compute second and higher order sidebands. Addition of these with the correct power weighting, together with the residual carrier tone, yields the required result. An example of a frequency modulated 300 channel radio relay system spectrum with pre-emphasis calculation by this method is shown in Figure 3.5, and a 960 channel system in Figure 3.6.

It is obvious that the length of the calculation depends on P^2 since as P^2 gets larger so more convolutions will have to be made to include the order of sidebands carrying the significant power. (Examination of the Poisson distribution shows that the greatest power is carried by those sidebands of order near the value of P^2). As the order of the sidebands increases, however, so the bandwidth

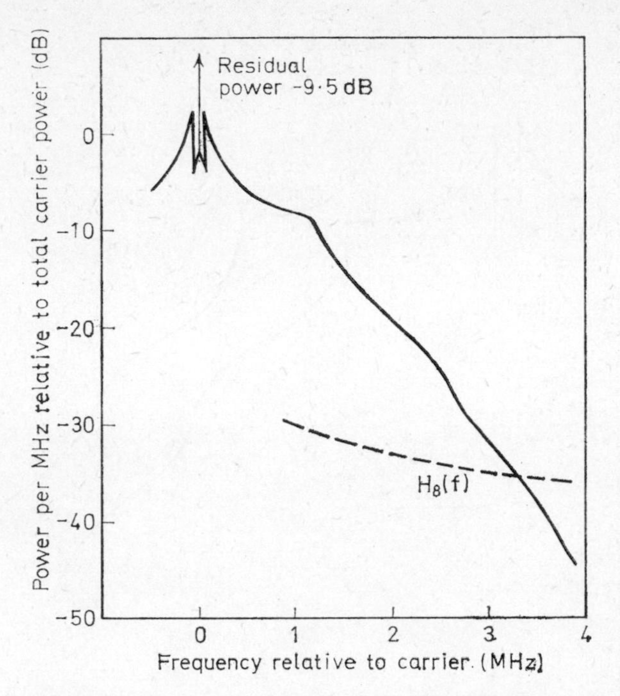

Figure 3.5 The spectrum of a 300 channel line-of-sight radio relay system. Minimum modulating frequency 0·064 MHz. *Maximum modulating frequency* 1·296 MHz. *Parameters and pre-emphasis CCIR recommendations 401, 404, 275*

they occupy increases and even though the total power may be significant, the contribution to the spectrum shape as a whole, especially at frequencies near to the carrier may not be significant. Thus the number of convolutions for accuracy over a given range is not related to P^2 in any simple way.

A way of testing over what range of frequencies a spectrum, calculated to a certain number of sidebands, is accurate, involves calculating the power contributed by the orders of sidebands omitted. If the spectrum is calculated to the $(x-1)$th order sidebands then the power Z_x contributed by the sidebands of order x to ∞ is given by

$$Z_x = \sum_{n=x}^{n=\infty} \frac{(P^2)^n}{n!} \tag{3.18}$$

It is now assumed that within the frequency range $-f$ to $+f$ relative to the carrier, the residual power Z_x of the sidebands of order x to ∞ is uniformly distributed giving a flat spectrum of height $H_x(f)$. The function $H_x(f)$ is plotted against f, and since the process of convolution tends to produce a Gaussian shape rather

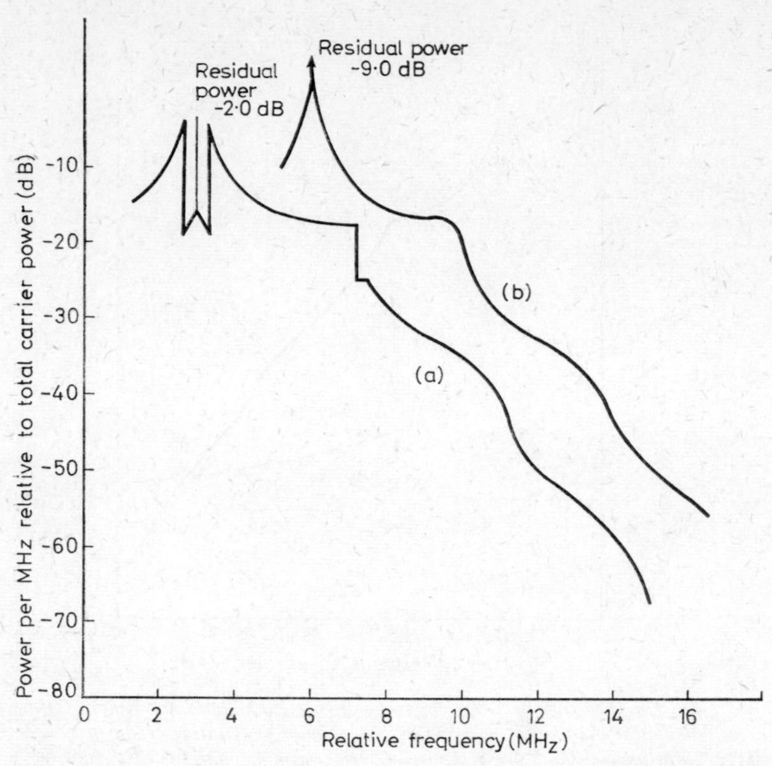

Figure 3.6. The spectra of 960 channel line-of-sight radio relay systems. (a) Minimum modulating frequency 0·316 MHz, *maximum modulating frequency* 4·188 MHz; *(b) minimum modulating frequency* 0·060 MHz, *maximum modulating frequency* 4·028 MHz. *Parameters and pre-emphasis CCIR recommendations 401, 404, 275*

than a flat shape, the spectrum of the sidebands of order x to ∞ must lie below this curve. $H_x(f)$ therefore provides an upper bound on the contributions of the sidebands of order x to ∞.

A curve of $H_8(f)$ ($Z_8 = 0{\cdot}002$) is plotted in Figure 3.5 where calculations are made to the 7th order sidebands and it can be seen that the spectrum is accurate to at least 3·2 MHz. (Note that the addition of one more order of sidebands alters the spectrum by less than 2 dB at 5·8 MHz and by 0·4 dB at 4 MHz).

Example 3.1

A baseband of telephone channels in frequency division multiplex extends from zero frequency to 3 MHz and the baseband is used to frequency modulated a microwave carrier. Pre-emphasis of the baseband power spectrum is employed and the characteristic is given by

$$pe(f) = (f/f_m)^2$$

The total mean square frequency deviation after pre-emphasis is 0·3 (MHz)^2.

Obtain an approximation for the shape of the spectrum about the carrier.

Solution

The baseband spectrum as applied to a frequency modulator is given by (see Figure 3.7a)

$$S_b(f) = \frac{k^2 f^2}{f_m^2} \quad \text{(MHz)}^2 \text{ per MHz}$$

for
$$-f_m < f < f_m$$
$$S_b(f) = 0 \quad \text{elsewhere}$$

Where k^2 = a constant of dimensions (MHz)^2 per MHz
f_m = top baseband frequency (3 MHz)

The mean square frequency deviation is

$$\Delta F^2 = \int_0^{f_m} \frac{2k^2 f^2}{f_m^2} \, df = 0{\cdot}3 \text{ (MHz)}^2$$

i.e.
$$\frac{2k^2 f_m}{3} = 0{\cdot}3$$

$$k^2 = 0{\cdot}15 \text{ (MHz)}^2 \text{ per MHz}$$

From equation 3.7 it can be seen that the equivalent baseband spectrum which would need to be applied to a phase modulator to give the same result is given by

$$S_\phi(f) = \frac{k^2 f^2}{f_m^2 f^2} = \frac{k^2}{f_m^2} \text{ (rad)}^2 \text{ per MHz}$$

Note that k^2/f_m^2 is a constant and therefore the spectrum applied to a phase modulator is rectangular as shown in Figure 3.7b.

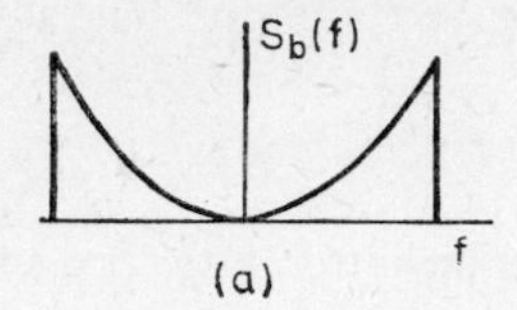

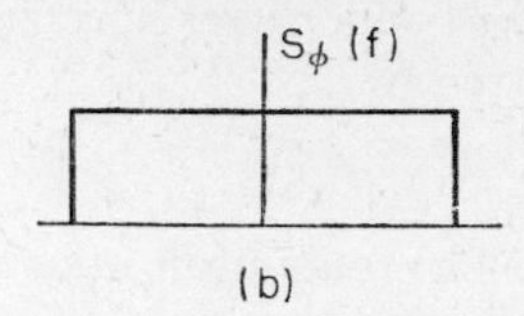

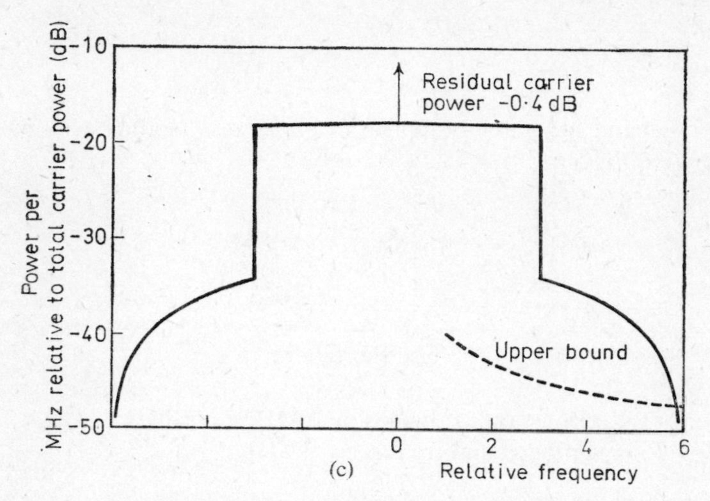

Figure 3.7 The spectra of Example 3.1. (a) Baseband spectrum for f.m.; (b) baseband spectrum for p.m.; (c) spectrum of the modulated signal about the carrier

The mean square phase deviation, P^2, is

$$P^2 = \int_0^{f_m} 2\frac{k^2}{f_m^2}\,\mathrm{d}f = \frac{2k^2}{f_m}\ (\text{rad})^2$$

i.e.

$$P^2 = 0{\cdot}1\ (\text{rad})^2$$

Since P^2 is small the formula of equation 3.17 can be used effectively because the first few terms will predominate and in fact the first three terms can be calculated quite simply.

The normalised baseband spectrum, $w_\phi(f)$ in equation 3.17 is given by

$$w_\phi(f) = \tfrac{1}{6} \qquad -f_m < f < f_m$$

The convolution of $w_\phi(f)$ with itself which is part of the third term of equation 3.17 is an isosceles triangle of height $\frac{1}{6}$ and base from -6 MHz to $+6$ MHz. Substituting for P^2 in equation 3.17,

$$G_s(f) = 0{\cdot}90484\,[\delta(f) + 0{\cdot}1\, w_\phi(f) + 0{\cdot}005\, w_\phi(f) \overset{2}{*} w_\phi(f) + \ldots]$$

$G_s(f)$ is plotted in Figure 3.7c, where it is to be noted that a logarithmic scale has been used as the ordinate.

Points to note

1. Since P^2 is small in this example most of the carrier power is at the carrier frequency (over 90% of the total power is in the delta function at the carrier frequency).

2. Most of the remaining power is in the first order sideband which is a replica of the baseband modulating signal without pre-emphasis. In fact the pre-emphasis characteristic in this problem was chosen not only to simplify the problem and enable the second order sidebands to be calculated, but also because this characteristic is ideal from the point of view of equalising the contribution from each channel to the r.f. power spectrum. By arranging things in this way the thermal noise falling in each channel is the same and no one channel is worse than another. In practice pre-emphasis characteristics are not ideal and do not attenuate the low frequency channels to the required amount. The result is that in practical systems the signal-to-thermal noise in the top baseband channel is usually worse than in the lower baseband channels.

3. The power spectrum shown in Figure 3.7c is calculated using the first three terms of equation 3.17, and the total power contribution of these terms is given by

$$\text{Power contribution of first three terms} = 0{\cdot}904\,84\,(1 + 0{\cdot}1 + 0{\cdot}005)$$

$$= 0{\cdot}999\,848$$

The remaining power which has been neglected is therefore about 0·0002 (or 0·02%). This is Z_3 used for the calculation of the upper bound.

This upper bound, calculated according to the proposals of Section 3.5.2 is shown dotted in Figure 3.7c which shows that for frequencies up to nearly 6 MHz either side of the carrier, the spectrum is accurately represented.

3.5.3. Method of Fluctuation Waves[5-7]

This method provides an alternative approach to the computation of the series of equation 3.17 based on a method originally proposed by Gladwin in 1944. The baseband is again represented by ordinates, i.e. delta functions having a power equal to the power contained in the width between the ordinates of the actual spectrum at each point. (These ordinates or delta functions are termed fluctuation waves by Gladwin). The spectrum of a single fluctuation wave

$$
e^{-P_k^2}\quad
\begin{array}{c} 1 \\ P_k^2/2 \\ (P_k^2/2)^2 \\ (P_k^2/2)^3 \\ (P_k^2/2)^4 \end{array}
\qquad
\begin{array}{ccccccccc}
 & & & & 1 & & & & \\
 & & & 1 & & 1 & & & \\
 & & 1 & & 2 & & 1 & & \\
 & 1 & & 3 & & 3 & & 1 & \\
1 & & 4 & & 6 & & 4 & & 1 \\
 & & & & & & & & \\
4 & 3 & 2 & 1 & 0 & 1 & 2 & 3 & 4
\end{array}
$$

multiples of f_k

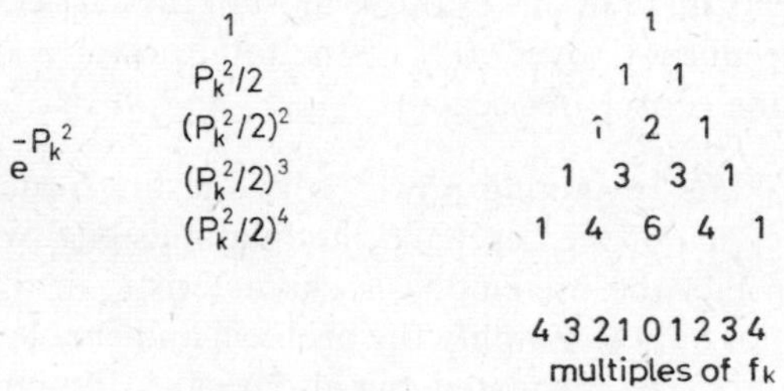

Figure 3.8 Convolution of fluctuation waves

in the baseband at frequency f_k is therefore represented by two delta functions of magnitude $P_k^2/2$ at frequencies $-f_k$ and $+f_k$, P_k^2 being the total mean square deviation of the fluctuation wave. The r.f. spectrum may then be obtained by summing the various order sidebands which for a single fluctuation wave or pair of delta functions are obtained by the simple process of convolution as shown in Figure 3.8. The residual carrier is $e^{-P_k^2}$ situated at zero frequency (with respect to the carrier) shown in the first row of Figure 3.8 and the first order sidebands are represented by the second row where the power is $e^{-P_k^2} \times P_k^2/2$ situated at frequencies $+f_k$ and $-f_k$. The nth order sidebands have the multiplier $(P_k^2/2)^n/n!$ and the delta functions are arranged as in the $(n+1)$th row of Pascal's triangle which results from the successive convolution of a spectrum of two delta functions with itself. Figure 3.8 shows that the power of each fluctuation wave at frequency nf_k in the $(n+2s)$th order sidebands is given by the coefficient of x^s in the expansion of $(1+x)^{n+2s}$ and its multiplier

$$\frac{(P_k^2/2)^{n+2s}}{(n+2s)!}$$

Adding all such components for $s = 0$ to ∞, the total power in the fluctuation waves at frequency nf_k is given by,

$$G_{sk} = e^{-P_k^2} \sum_{s=0}^{\infty} {}^{n+2s}C_s \frac{(P_k^2/2)^{n+2s}}{(n+2s)!}$$

$$= e^{-P_k^2} \sum_{s=0}^{\infty} \frac{(n+2s)(n+2s-1)\ldots(n+s+1)}{s!\,(n+2s)!} (P_k^2/2)^{n+2s}$$

$$= e^{-P_k^2} \sum_{s=0}^{\infty} \frac{1}{s!\,(n+s)!} (P_k^2/2)^{n+2s} \quad (3.19)$$

$$= e^{-P_k^2}\, j^{-n} J_n(jP_k^2)$$

$$= e^{-P_k^2}\, I_n(P_k^2) \quad (3.20)$$

The last three equations are standard relationships for Bessel and modified Bessel functions (see Appendix 5) showing how the results agree with those obtained by Gladwin. Thus a simple numerical method can be devised for the calculation of the spectrum of a fluctuation wave.

In practice the baseband will be represented by a number of fluctuation waves and the results of Appendix 3 concerning the addition of signals in the baseband can be used to calculate the total r.f. spectrum. Thus having obtained the separate r.f. spectra of all the fluctuation waves, these are then numerically convolved with each other to give the final spectrum.

3.5.4. Method using Baseband Spectra with Minimum Modulating Frequency Extending to Zero[8, 9]

This method provides a solution to equation 3.14

$$G_s(f) = \int_{-\infty}^{\infty} \exp[-R(0)+R(\tau)]\, e^{-j\omega\tau}\, d\tau$$

by evaluating the autocorrelation function for the baseband. It is then possible to expand the expression for the autocorrelation function in powers of the minimum modulating frequency (f_0), and the first term can be extracted. This corresponds to the baseband extending to zero frequency (i.e. $f_0 = 0$). The details of this technique are given in the references and are not discussed here. The value of the results lies in the fact that generalised power spectra can be obtained since, for $f_0 = 0$, a family of spectra can be plotted on axes of power per unit bandwidth$\times f_m$ as ordinate and f/f_m as abscissa[8]. It should be noted that these spectra cannot be obtained from equation 3.14 by the method of successive convolution or by the method of fluctuation waves since P^2 tends to infinity as f_0 tends to zero.

The usefulness of the method for zero minimum baseband modulating frequency lies in the fact that the r.f. spectrum for realistic values of f_0 differs very little from the spectrum for $f_0 = 0$. This can be seen by visualising the f_0-f_m r.f. spectrum and using the addition theorem of Appendix 3 for f.m. spectra to obtain the $0-f_m$ spectrum. Thus to obtain the $0-f_m$ r.f. spectrum the f_0-f_m r.f. spectrum is convolved with an r.f. spectrum calculated for a baseband extending from $0-f_0$. This latter spectrum, for small values of f_0, will be narrow compared with the other two and so the main effects of the convolution will be to iron out the discontinuities in the f_0-f_m spectrum. In particular the residual carrier and discontinuities at $f=f_0$ are smoothed out and the discontinuity at $f=f_m$ is smoothed out. The spectrum of a 960 channel radio relay system calculated by the method of successive convolution for $f_\theta = 0{\cdot}316$ MHz and $f_0 = 0{\cdot}060$ MHz shown in Figure 3.6 illustrates this smoothing out process.

3.5.5. The Quasistationary Approach[10, 11]

The quasistationary approach to calculation of the r.f. spectrum is based on the argument that if the carrier frequency varies very slowly, then the amount of power falling within any particular frequency slot in the spectrum is proportional to the amount of time the carrier remains in that slot. In terms of a random modulating signal, this means that the power spectrum at any frequency is proportional to the probability of finding the carrier deviated to that frequency and the spectrum takes on the shape of the amplitude probability distribution of the modulating signal.

Since the carrier frequency is required to vary very slowly a criterion for the quasistationary argument to be applicable is that the maximum modulating frequency of the baseband should be low compared with the deviation of the carrier, i.e. the ratio $\Delta F/f_m$ should be large. It is found in practice that results are reasonably accurate for $\Delta F/f_m$ greater than one or two. There are other factors which affect the applicability such as the ratio f_m/f_0, which must be also large, and therefore it is difficult to give rules for any particular accuracy of results.

In general communication satellite systems modulated with a f.d.m. speech channel baseband with a reasonable degree of loading have parameters which meet those required for the quasistationary argument.

Example 3.2

A communication satellite system carries 120 channels of telephony in a baseband extending from 0·01 MHz to 0·5 MHz. The baseband frequency modulates a carrier with a total mean square frequency deviation of 2·56 MHz. Calculate the r.f. frequency spectrum.

Solution

In this problem f_m/f_0 is large. $\Delta F = 1{\cdot}6$ MHz and so the ratio $\Delta F_m/f$ is 3·2 which is also large enough for the quasistationary argument to apply. The amplitude probability density of the baseband will be Gaussian and therefore the spectrum can be written down as

$$G_s(f) = \frac{1}{\sqrt{(2\pi)}1{\cdot}6}\, e^{-f^2/(2{\cdot}56)2}$$

3.5.6. Monte Carlo Method[12]

In this computer method the baseband is represented by a large number of sine waves of incrementally increasing frequency, random phase angles and whose amplitudes correspond to the spectral height of the baseband. This is a standard way to represent a noise signal as indicated in Section 2.2 and the closeness of the approximation of the amplitude probability distribution to a Gaussian distribution depends on the number of sine waves taken, and on the number of times the process is repeated with different sets of random phase angles.

The shape of certain frequency modulation spectra are controlled to a large degree by the shape of the baseband spectra and the latter can be represented accurately by one set of sine waves of incrementally increasing frequency and random phase angles. Thus calculations for only a few sets of random phase angles for the sine waves will result in good convergence on the final frequency modulation spectrum in this case.

The shape of other frequency modulation spectra, however, depend to a large degree on the amplitude probability distribution of the baseband signal. In this case calculations must be made for a large number of sets of random phase angles as indicated earlier and the Monte Carlo method may not be so suitable in this case.

The criteria for the shape of the frequency modulation spectrum are shown by referring back to equation 3.17 where it may be seen that if P^2 is small and therefore ΔF^2 is small the first few terms predominate and the frequency modulation spectrum is controlled by the baseband shape $w_\phi(f)$. The arguments contained in Section 3.5.4 can be used to show further that it is only ΔF^2 that is required to be small with respect to f_m i.e. $\Delta F^2/f_m$ small for these conditions to apply. This follows since extension of f_0 towards zero (resulting in a large value of P^2) does not affect the f.m. spectrum shape appreciably.

The quasistationary argument on the other hand (Section 3.5.5) indicates that for a system with large $\Delta F^2/f_m$ the f.m. spectral shape is independent of the baseband shape and depends only on the amplitude probability distribution of the baseband.

Calculation of spectra is only an intermediate step in the calculation of distortion in systems for which the Monte Carlo method is very useful (See Chapter 5).

3.5.7. Direct Method using Fast Fourier Transforms[17]

In this method the autocorrelation function of the baseband $(R_\phi(\tau))$ is found by using numerical methods of calculating the Fast Fourier transform. Equation 3.13 is then formed by taking the exponential as follows,

$$R_s(\tau) = \exp\left[-R_\phi(0)+R_\phi(\tau)\right]$$

Finally the power spectrum of $s(t)$ (the required output spectrum) is found by using numerical methods again for the Fourier transform.

This method has been found useful for calculating spectra for large values of P^2 and ΔF^2 where the quasistationary approach is questionable.

3.6. INTERFERENCE INTO AMPLITUDE MODULATED SYSTEMS

3.6.1. Interference into Sychronous Detectors (s.s.b.)

The baseband modulating signal is assumed to have a noise-like symmetrical power spectrum stretching from $-f_m$ through 0 to $+f_m$, the process of single sideband modulation assumes that all the necessary information in this spectrum is contained between 0 and

$\pm f_m$ and shifts one half of the spectrum up to the r.f. frequency required as can be seen from Figure 3.3. The process of synchronous detection is merely to translate the single sideband back to zero frequency and this is achieved by multiplying the single sideband by an exact replica of the carrier frequency. (This is often generated from the suppressed carrier which is transmitted with the signal).

Thus if the single sideband signal represented by the function $f_w(t) \cos \omega_c t$ has an interfering signal $f_u(t) \cos [\omega_c t+\phi(t)+\omega_D t]$ then the output of the demodulator is given by

$$f_w(t) \cos \omega_c t \cos \omega_c t+f_u(t) \cos [\omega_c t+\phi(t)+\omega_D t] \cos \omega_c t \quad (3.21)$$

The spectrum of the noise appearing in the baseband due to interference is therefore given by the convolution of the spectrum of the interfering signal with the spectrum of $\cos \omega_c t$ (This follows since the process of multiplication in the time domain is transformed to convolution in the frequency domain—see Appendix 1).

In the interference calculations that follow the spectrum about the r.f. carrier is taken as normalised to unit total power, while the true spectrum has one half of the power in the positive frequency plane and one half in the negative half plane.

Thus if the spectrum of the interference about its carrier is $G_u(f)$, then the spectrum of the interference centred on the frequency f_c+f_D may be written as

$$\tfrac{1}{2} G_u[f-(f_c+f_D)]+\tfrac{1}{2} G_u[f+(f_c+f_D)]$$

If this is convolved with the spectrum of $\sqrt{2} \cos \omega_c t$ (which has unit total power) then the result shown in Figure 3.9 is

$$\begin{aligned} &\tfrac{1}{4} G_u(f-f_D)+\tfrac{1}{4} G_u(f+f_D)+\tfrac{1}{4} G_u[f-(2f_c+f_D)] \\ &\quad +\tfrac{1}{4} G_u[f+(2f_c+f_D)] \end{aligned} \quad (3.22)$$

Only the first two terms appear at baseband frequencies and so the second two are ignored. A similar treatment of the signal single sidebands produces the wanted baseband double sided spectrum $\frac{1}{4} G_w(f)$ where again the r.f. signal is taken as normalised to unit power.

From Figure 3.9, the interference transfer factor, as defined in Section 3.1, for a narrow frequency slot centred on the frequency α is given by

$$X = \frac{G_w(\alpha)}{G_u(\alpha-f_D)+G_u(\alpha+f_D)} \quad (3.23)$$

The above expression ignores the action of any r.f. filtering in the receiver, and if this were such as to allow the single sideband signal

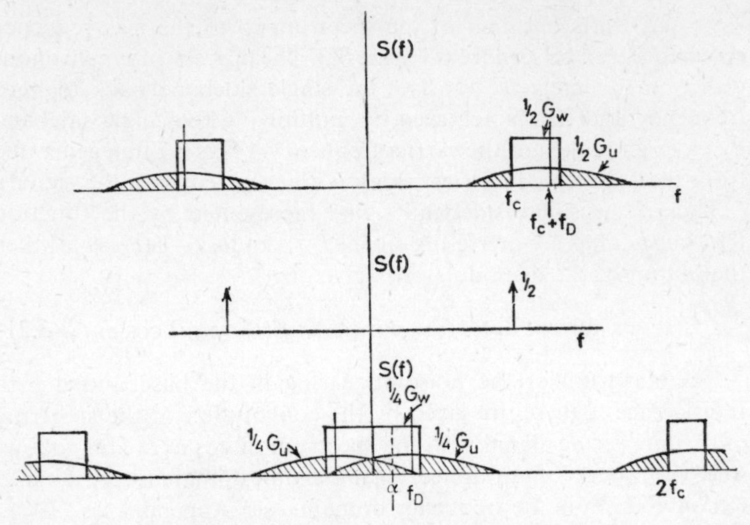

Figure 3.9 Interference into s.s.b.-s.c. systems

to pass through while eliminating interference at other frequencies (shaded in Figure 3.9) then it can be seen that X becomes

$$X = \frac{G_w(\alpha)}{G_u(\alpha - f_D)} \tag{3.24}$$

3.6.2. Interference into Envelope Detectors

Envelope detectors are generally used with double sideband amplitude modulation, and as shown in Section 3.4.2 this type of modulation can be represented by the equation,

$$\begin{aligned} s(t) &= [1+f(t)] \cos \omega_c t \\ &= f_w(t) \cos \omega_c t \end{aligned}$$

where $f_w(t)$ is put equal to $1+f(t)$ for convenience.

The interference added to the signal gives, for the total input to the receiver demodulator, (equation 3.1)

$$f_w(t) \cos \omega_c t + f_u(t) \cos [\omega_c t + \omega_D t + \phi(t)]$$

and from equation 3.3, the amplitude of this expression is given by,

$$A^2 = f_w^2(t) + 2 f_w(t) f_u(t) \cos [\omega_D + \phi(t)] + f_u^2(t) \tag{3.25}$$

If it is assumed that the total noise power is small compared with the signal term, the last term in equation 3.25 can be ignored and

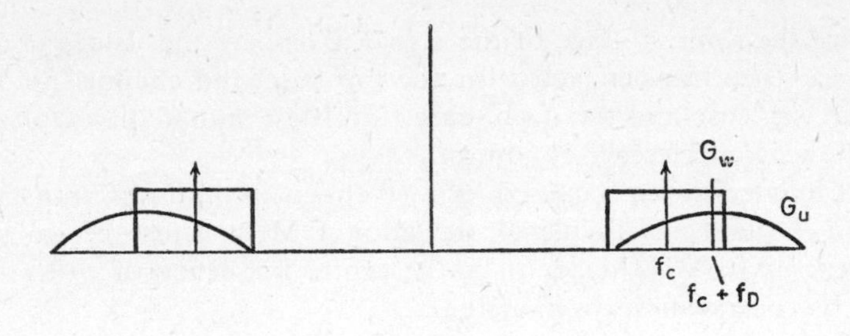

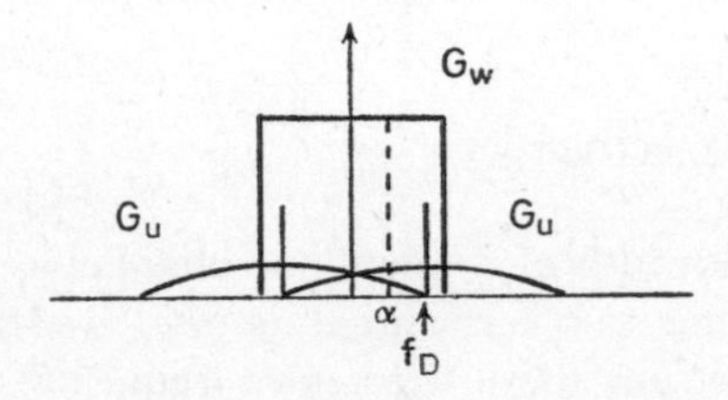

Figure 3.10 Interference into d.s.b. systems

thus

$$A^2 = f_w^2(t)\left\{1+\frac{2f_u(t)\cos[\omega_D t+\phi(t)]}{f_w(t)}\right\}$$

which for small $f_u(t)/f_w(t)$ gives

$$A = f_w + f_u(t)\cos[\omega_D t+\phi(t)]$$

For small noise inputs, the output of an envelope detector consists of the original signal plus the interference noise spectrum translated from a centre frequency of f_c+f_D to f_D. Let $G_w(f)$ be the normalised double sided spectrum of $f_w(t)$ and $G_u(f)$ be the normalised spectrum of the interference about its centre frequency as in Section 3.6.1. For a narrow frequency slot centred on the frequency α, the interference transfer factor as defined in Section 3.1 is given from Figure 3.10 as

$$X = \frac{G_w(\alpha)}{\frac{1}{4}G_u(f_D-\alpha)+\frac{1}{4}G_u(f_D+\alpha)} \qquad f_D \neq 0 \qquad (3.26)$$

Figure 3.10 shows that no amount of careful r.f. filtering will improve this figure.

Example 3.3

500 channels of telephony in frequency division multiplex are to be transmitted using either s.s.b.-s.c. modulation or d.s.b. modulation, the voltage peaks of the baseband signal being clipped at

twice the r.m.s. voltage of the signal. Compare the values of the interference transfer factor for the top baseband channel for the s.s.b.-s.c. case and the d.s.b. case (for 100% modulation) for the following interference condition.

The interference is caused by a 60 channel f.d.m.-f.m. transmission of total r.m.s. frequency deviation 1 MHz whose centre frequency is 0·5MHz higher than the centre frequency of either the s.s.b.-s.c. system or d.s.b. system.

Solution

Wanted Signal Spectrum s.s.b.-s.c.

Taking the bandwidth of a telephone channel as 4 kHz, the total bandwidth of the s.s.b.-s.c. signal is 500×4 kHz = 2MHz. The normalised spectrum about the centre frequency is then

$$G_w(f) = \tfrac{1}{2}$$
$$\text{s.s.b.-s.c.} \qquad -1\text{ MHz} < f < +1\text{ MHz}$$

Wanted Signal Spectrum d.s.b.

In this case if the clipping voltage is h then for 100% modulation the modulating signal must be given a d.c. bias of h volts in order that the modulating signal is never negative as shown in Figure 3.11.

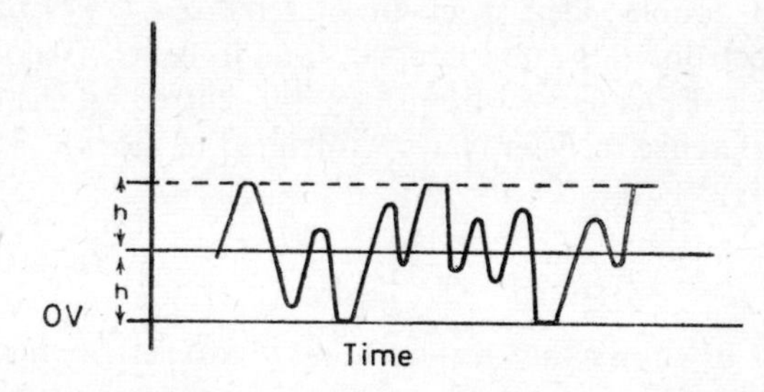

Figure 3.11 Clipping of the baseband for d.s.b. modulation

If the r.m.s. voltage of the signal is v volts then the ratio of the power in the signal to the total modulating power is

$$\text{signal power ratio} = \frac{v^2}{h^2+v^2}$$

The ratio of the power at the carrier frequency to the total power is

$$\text{carrier power ratio} = \frac{h^2}{h^2+v^2}$$

Thus the normalised spectrum of the wanted signal consists of a carrier component of strength $\frac{h^2}{h^2+v^2}$ and two sideband components each of 2 MHz bandwidth and each of total power $\frac{v^2}{2(h^2+v^2)}$.

In this example $h/v = 2$ and so the spectrum about the carrier becomes

$$G_w(f) = \tfrac{1}{20} + \tfrac{4}{5}\,\delta(f)$$

d.s.b. $\quad -2\text{ MHz} < f < 2\text{ MHz}$

Interference Spectrum

The maximum modulating frequency (f_m) for the interfering system is 4×60 kHz $= 0{\cdot}24$ MHz and the ratio of the r.m.s. frequency deviation (ΔF) to f_m is 4·2 and the system satisfies the requirements for the quasistationary method for f.m. spectra of Section 3.5.5. The spectrum is therefore

$$G_u(f) = \frac{1}{\sqrt{(2\pi)}}\,e^{-f^2/2}$$

Calculation of X for s.s.b.-s.c. System

From equation 3.24

$$X = \frac{G_w(\alpha)}{G_u(\alpha - f_D)}$$

For the top baseband channel $\alpha = 1$ MHz (from centre frequency) and $(\alpha - f_D) = 0{\cdot}5$ MHz

$$G_w(\alpha) = 0{\cdot}5$$

and

$$G_u(\alpha - f_D) = 0{\cdot}35$$

$$X = \frac{0{\cdot}5}{0{\cdot}35} = 1{\cdot}43$$

i.e.

$$X = 1{\cdot}6\text{ dB}$$

Calculation of X for d.s.b. System

In equation 3.26

$$\alpha = 2\text{ MHz} \quad G_w(\alpha) = \tfrac{1}{20} = 0{\cdot}05$$

$$(f_D - \alpha) = -1{\cdot}5\text{ MHz} \quad G_u(f_D - \alpha) = 0{\cdot}13$$

$$(f_D + \alpha) = 2{\cdot}5\text{ MHz} \quad G_u(f_D + \alpha) = 0{\cdot}02$$

$$X = \frac{0{\cdot}05 \times 4}{0{\cdot}15} = 1{\cdot}33$$

i.e.

$$X = 1{\cdot}2\text{ dB}$$

Points to Note

1. Careful examination of the above example will show that the value of X is critically dependent on the value of α and f_D and it cannot be concluded that d.s.b. systems perform in approximately the same way as s.s.b.-s.c. systems under interference conditions. If f_D for the s.s.b.-s.c. system had been interpreted as the difference in frequency between the lower end of the spectrum and the interference carrier, then X would have been considerably larger.

2. For wideband interference such as that obtained for thermal noise, the results for the s.s.b.-s.c. system are much more favourable compared to the d.s.b. system for all values of α.

Taking the case of a flat-topped thermal noise spectrum and a flat-topped wanted signal, for the s.s.b.-s.c. system $G_w(f)$ and $G_u(f)$ are identical and the transfer factor is unity.

i.e.

$$X_{\text{s.s.b.-s.c.}} = 1$$

For the case of d.s.b., if it is assumed that the modulating signal $f_w(t)$ consists of the d.c. bias and a flat topped signal spectrum, and if the total double sided bandwidth is B, then the normalised spectrum of the d.s.b. is given (as shown in the example) by

$$G_w(f) = \frac{h^2}{h^2+v^2}\delta(f) + \frac{v^2}{B(h^2+v^2)} \quad \text{(over the bandwidth } B\text{)}$$

The noise power density spectrum is also assumed to be flat and for the same bandwidth its normalised spectrum becomes

$$G_u(f) = \frac{1}{B}$$

and also since the spectrum is flat

$$G_u(f_D-\alpha) = G_u(f_D+\alpha)$$

Thus for values of α other than zero equation 3.26 gives

$$X_{\text{d.s.b.}} = \frac{2v^2}{(h^2+v^2)}$$

It should be noted that this result implicitly assumes that $f_D \neq 0$. Since wideband noise is not centered around any obvious 'carrier' it could be argued that f_D might just as well be taken as zero. This however would imply that there was correlation between frequency components in the noise spectrum symmetrically displaced about the wanted system carrier, as in d.s.b. spectra. The noise therefore is misrepresented in this case, but for $f_D \neq 0$, there is no such symmetry about the carrier of the wanted system and the interference 'looks' like noise.

There is a more rigorous approach to this problem using the representation of narrow band noise given in Section 2.2. Example 2.1 (equation 2.11) shows that narrowband noise can be written as

$$R(t) \sin [\omega_c t+\phi(t)] = x_c(t) \cos \omega_c t+x_s(t) \sin \omega_c t$$

where $x_c(t)$ and $x_t(t)$ are Gaussian random variables each of power equal to the narrow band noise.

Thus the signal plus noise to the input of the envelope detector may be written as

$$\begin{aligned} &f_w(t) \cos \omega_c t+x_c(t) \cos \omega_c t+x_s(t) \sin \omega_c t \\ &= [f_w(t)+x_c(t)] \cos \omega_c t+x_s(t) \sin \omega_c t \end{aligned}$$

If x_c and x_s are small compared to f_w then the quadrature component x_s can be neglected, and the output from the envelope detector is the envelope of the first term and is

$$A = f_w(t)+x_c(t)$$

If the power in $f_w(t)$ is P_{in} and the noise power at the input is N_{in}, then the output signal/noise ratio is given by

$$\frac{P_{\text{out}}}{N_{\text{out}}} = \frac{2P_{\text{in}}}{N_{\text{in}}}$$

Since the signal power in $f_w(t)$ is $v^2/(h^2+v^2)$ of the total, then the transfer factor is

$$X = \frac{2v^2}{h^2+v^2}$$

3.7. INTERFERENCE INTO ANGLE MODULATED SYSTEMS[3,13,14]

3.7.1. General Formulae

In angle modulated systems the wanted signal has a constant amplitude and varying angle and is represented by

$$s(t) = A_w \cos [\omega_c t + \phi_w(t)]$$

The expression for the sum of the wanted signal and the unwanted interference (Equation 3.1) then becomes

$$s(t) = A_w \cos [\omega_c t + \phi_w(t)] + f_u(t) \cos [\omega_c t + W_D t + \phi_u(t)] \tag{3.27}$$

Angle modulation receivers generally have amplitude limiters incorporated to suppress amplitude variations and so the resultant amplitude of equation 3.27 which is given by equation 3.3 is treated as constant and taken as unity for convenience. Thus equations 3.2 and 3.3 give

$$s(t) = \cos [\omega_c t + \phi_w(t) + \mu(t)] \tag{3.28}$$

It can be seen that the demodulator will treat the incoming signal as a carrier phase modulated by the wanted signal and additionally disturbed in phase by the interference $\mu(t)$. From equation 3.4

$$\mu(t) = \frac{f_u(t) \sin [\omega_D t + \phi_u(t) - \phi_w(t)]}{A_w + f_u(t) \cos [\omega_D t + \phi_u(t) - \phi_w(t)]} \tag{3.29}$$

If the amplitude of the interference $f_u(t)$ is small compared with the amplitude of the wanted r.f. signal A_w then $\mu(t)$ may be written as

$$\mu(t) = \frac{f_u(t)}{A_w} \sin [\omega_D t + \phi_u(t) - \phi_w(t)] \tag{3.30}$$

From equation 3.28 the demodulated signal is $[\phi_w(t) + \mu(t)]$ and so to calculate the interference transfer factor it is necessary to devise an expression for the power spectrum $S_\mu(f)$ of $\mu(t)$ and then superimpose this on the spectrum $S_\phi(f)$ of the modulating signal $\phi_w(t)$.

It is shown in Appendix 3 that the spectrum about a carrier which is modulated by the sum of two statistically independent phase disturbances is given by the convolution of the spectra about two individual carriers modulated by each phase disturbance separately. Thus, if the normalised spectrum about the carrier ω_1 of $\sin[\omega_1 t+\phi_w(t)]$ is $G_w(f)$ and the normalised spectrum about the carrier frequency ω_2 of $\sin[\omega_2 t+\phi_u(t)]$ is $G_u(f)$ then the normalised spectrum $G_{u+w}(f)$ about the carrier ω_3 of $\sin[\omega_3 t+\phi_w(t)+\phi_u(t)]$ is given by

$$G_{u+w}(f) = G_w(f) * G_u(f) \tag{3.31}$$

Appendix 3 also shows that this result assumes that the frequency ω_3 is large compared with the spectral width of $G_{u+w}(f)$ and that if this is not the case when the skirts of $G_{u+w}(f)$ from the negative frequency plane may overlap into the positive frequency plane and

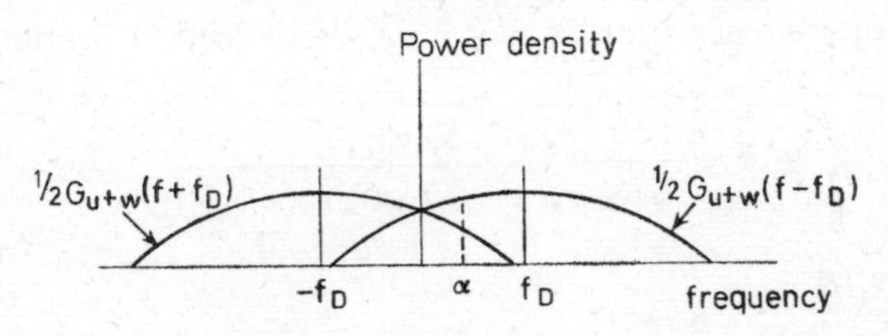

Figure 3.12 Overlap of interference spectra in angle modulation

vice-versa, as shown in Figure 3.12. Thus the normalised spectrum of $\sin[\omega_D t+\phi_u(t)+\phi_w(t)]$ which is the frequency modulated portion of $\mu(t)$ is given by

$$\tfrac{1}{2}[G_{u+w}(f-f_D)+G_{u+w}(f+f_D)]$$

The total power in $\sin[\omega_D t+\phi_u(t)+\phi_w(t)]$ is $\frac{1}{2}$ and the power of the ratio $\dfrac{f_u(t)}{A_w}$ is simply the ratio of the power in the r.f. interference to the power of the wanted r.f. signal N_{in}/S_{in}. Since these functions are multiplied together to give $\mu(t)$ and since from Appendix 1 the process of multiplication is transformed to convolution, $G_u(f)$ may be taken to be the complete normalised spectrum of the interfering signal. The spectrum of $\mu(t)$ is thus given by

$$S_\mu(f) = \frac{1}{4}\frac{N_{in}}{S_{in}}[G_{u+w}(f-f_D)+G_{u+w}(f+f_D)] \tag{3.32}$$

$$G_{u+w}(f) = G_u(f) * G_w(f) \tag{3.33}$$

where

$\dfrac{S_{in}}{N_{in}}$ = the ratio of power in the wanted r.f. signal to the power in the interference r.f. signal.

$G_w(f)$ = the normalised spectrum of the wanted r.f. signal about its carrier.

$G_u(f)$ = the spectrum of the interfering r.f. signal about its carrier.

The S/N in a channel centred on frequency α is therefore given by

$$S/N = \frac{S_\phi(\alpha)}{S_\mu(\alpha)} \tag{3.34}$$

where

$S_\phi(f)$ = the normalised double sided spectrum of the baseband signal which phase modulates the carrier of the wanted system.

The interference transfer factor (X) as defined in Section 3.1 is thus given by

$$X = \frac{S_\phi(\alpha)}{\frac{1}{4}[G_{u+w}(\alpha - f_D) + G_{u+w}(\alpha + f_D)]} \tag{3.35}$$

Example 3.4

A line-of-sight radio relay link transmits a large number of telephone channels by frequency modulation of a baseband of the channels assembled in frequency division multiplex. The signal/noise ratio in any channel is defined as the ratio of a 1 mW test tone to interference power level in that channel. Derive a formula for the interference transfer factor (X) for the channel in terms of the following parameters:

S_{out}/N_{out} = Signal/noise ratio in a telephone channel as defined above.

X = Interference transfer factor defined by $S_{out}/N_{out} = X \times S_{in}/N_{in}$.

S_{in}/N_{in} = Ratio of power of wanted r.f. signal (frequency modulated) to power of the interfering r.f. signal.

α = Frequency of the channel in the baseband of the radio relay signal in which the interference is considered.

f_B = Bandwidth of the telephone channel in which the interference is being considered.

f_d = r.m.s. frequency deviation of the carrier by a single 1 mW test tone in the baseband, in the absence of pre-emphasis. (Note that this is independent of the position in the baseband of the tone).

$pe(f)$ = The pre-emphasis characteristic used (see Section 3.3).

f_D = The frequency separation between the carrier of the radio relay system and the carrier frequency or centre frequency of the interference.

Solution

From Equation 3.35 the interference transfer factor is given by

$$X = \frac{S_\phi(\alpha)}{\frac{1}{4}[G_{u+w}(\alpha - f_D) + G_{u+w}(\alpha + f_D)]} \tag{3.36}$$

$S_\phi(f)$ is the normalised double sided spectrum of the baseband which phase modulates the carrier of the radio relay system.

If there are assumed to be N channels of telephony each of bandwidth f_B then the bandwidth of the baseband $(f_m - f_0)$ is Nf_B. If each channel has a test tone of mean square frequency deviation f_d^2 then the total mean square deviation given to carrier would be Nf_d^2. Substituting in equation 3.7 for $S_\phi(f)$ which is the equivalent spectrum of the baseband as applied to a phase modulator

$$S_\phi(f) = \frac{\Delta F^2}{2(f_m - f_0)f^2} = \frac{Nf_d^2}{2Nf_B f^2}$$

$$S_\phi(f) = \frac{f_d^2}{2f^2 f_B} \tag{3.37}$$

If pre-emphasis is used then the baseband spectrum is modified to

$$S_\phi(f) = \frac{f_d^2 pe(f)}{2f^2 f_B} \tag{3.38}$$

Substituting equation 3.38 in equation 3.36,

$$X = \frac{\dfrac{f_d^2 p\, e(\alpha)}{2\alpha^2 f_B}}{\frac{1}{4}[G_{u+w}(\alpha - f_D) + G_{u+w}(\alpha + f_D)]}$$

i.e.

$$\frac{1}{X} = \frac{\alpha^2 f_B}{2f_d^2 pe(\alpha)}[G_{u+w}(\alpha - f_D) + G_{u+w}(\alpha + f_D)] \tag{3.39}$$

Example 3.5

Derive a formula for the signal/noise ratio in the baseband of a frequency modulated frequency division multiplexed system where the noise at the input to the demodulator is thermal noise with a flat spectrum.

Solution

The following parameters will be used,

S_{out}/N_{out} = Signal/noise ratio defined as the ratio of a 1 mW test tone to thermal interference level in a telephone channel.

B = The bandwidth of the frequency modulation receiver.

S_{in}/N_{in} = Ratio of the power of the frequency modulated signal to the thermal noise power at the demodulator terminals.

f_d = r.m.s. frequency deviation of the carrier by a single 1 mW test tone in the baseband.

α = Frequency of the channel in the baseband in which the interference is being considered.

f_B = Bandwidth of the telephone channel in which the interference is being considered.

$pe(f)$ = Pre-emphasis characteristic.

The interference transfer factor for such a system is given by equation 3.39 as

$$\frac{1}{X} = \frac{\alpha^2 f_B}{2f_d^2 pe(\alpha)} [G_{u+w}(\alpha - f_D) + G_{u+w}(\alpha + f_D)]$$

The problem is to find $G_{u+w}(f)$ which is given by (equation 3.33)

$$G_{u+w}(f) = G_u(f) * G_w(f)$$

In this case $G_w(f)$ is the spectrum of the frequency modulated system and $G_u(f)$ is the spectrum of the thermal noise. The latter is a rectangle of height $1/B$ and bandwidth B (the spectrum is normalised). Since the rectangle will be wide compared with the significant portion of $G_w(f)$, the convolution $G_w(f) * G_u(f)$ will also be a rectangle of height $1/B$.

i.e. $\quad G_{u+w}(f) = 1/B$

and

$$\frac{1}{X} = \frac{\alpha^2 f_B}{2f_d^2 p(\alpha)} [1/B + 1/B]$$

$$\frac{1}{X} = \frac{\alpha^2}{f_d^2} \frac{f_B}{B} \frac{1}{pe(\alpha)} \tag{3.40}$$

3.7.2. The Effect of Channel Occupancy on Interference

In many cases, the worst possible interference between two frequency modulated systems is associated with the delta functions in their spectra due to residual carrier power at the carrier frequency (see Example 3.1). In the case of narrow deviation systems (trunk telephony radio relay systems for example) this residual carrier power remains significant even under conditions of full channel occupancy. For instance, the residual carrier power in the system of Example 3.1 accounts for about 90% of the total carrier power and a similar situation occurs in practical line-of-sight trunk telephony radio relay systems. In the case of wide deviation systems (communication satellite systems for example) the residual carrier is negligibly small under condition of full loading. (Note that the residual carrier power is assumed to be zero in the quasistationary approach of Section 3.5.5.) Thus in this latter case, the interference situation under conditions of full loading is very much more favourable than under conditions of light loading where a high residual carrier will appear.

Example 3.6

Compare the interference transfer factor for interference into a lightly loaded 1800 channel radio relay system from (a) a 240 channel satellite system fully loaded (b) the same system lightly loaded. The r.m.s. deviation (ΔF) of the satellite system is 2 MHz, the carrier separation between the systems (f_D) is 8 MHz and the comparison is to be made for a channel at a frequency (α) of 8 MHz in the radio relay system.

Solution

In both cases the radio relay system is lightly loaded and so for convenience it will be assumed that its spectrum is simply a delta function at the carrier frequency. (The results would not be very much different if the radio relay system were taken to be fully loaded).

(a) Satellite System Fully Loaded

In equations 3.33 and 3.35 the above conditions give

$$G_w(f) = \delta(f)$$

$$G_u(f) = \frac{1}{\sqrt{(2\pi)2}}\,e^{-f^2/2\times 4}$$

using the quasistationary method for wide deviation systems.

$$G_{u+w}(f) = G_u(f) * G_w(f)$$

$$G_{u+w}(f) = \frac{1}{2\sqrt{(2\pi)}}\,e^{-f^2/8}$$

For $\alpha = 8$ MHz and $f_D = 8$ MHz

$$G_{u+w}(\alpha - f_D) = G_{u+w}(0) = 0{\cdot}2$$

$$G_{u+w}(\alpha + f_D) = G_{u+w}(16) = \text{negligible compared with } G_{u+w}(0)$$

$$G_{u+w}(\alpha - f_D) + G_{u+w}(\alpha + f_D) = 0{\cdot}2 \text{ power per MHz}$$

(b) Satellite System Lightly Loaded

In this case the spectrum of the satellite system will also be taken as a delta function at the carrier frequency, and so

$$G_w(f) = G_u(f) = G_{u+w}(f) = \delta(f)$$

$$G(\alpha - f_D) = \infty \qquad \text{(unit total power)}$$

$$G(\alpha + f_D) = 0$$

Since unit power falls into a channel at a frequency of 8 MHz in the baseband of the radio relay system the power density can be considered as $1/f_B$ where f_B is the channel bandwidth. Thus for a channel bandwidth of 4 kHz the equivalent power density becomes

$$G(\alpha - f_D) = \frac{1}{0{\cdot}004} \text{ power per MHz.}$$

$$= 250 \qquad \text{power per MHz.}$$

In equation 3.35 $S_\phi(\alpha)$ is the same in both cases a and b and so the ratio of the interference transfer factors for the two cases is

$$\frac{X\,(\text{satellite system loaded})}{X\,(\text{satellite system unloaded})} = \frac{250}{0{\cdot}2} = 1\,250$$

A similar calculation for the radio relay system fully loaded gives a ratio of 1 280.

The difference of about 31 dB in the interference level for a satellite system lightly loaded and a satellite system fully loaded shown in the above examples illustrates the importance of channel occupancy. In order to avoid the condition of light load in satellite systems carrier energy dispersal may be used. This involves the application of a low frequency waveform to the baseband in the region below the lowest frequency of the telephone channels. This waveform keeps the carrier moving in frequency and thereby spreads the energy over a wide bandwidth in a similar way to the telephone channels when the system is fully loaded. The applied waveform may be sinusoidal, sawtooth or a band of low frequency noise. The spectrum of the lightly loaded but dispersed system may be calculated using the quasistationary approach of Section 3.5.5 since the modulation waveform varies slowly. The theorems of Appendix 1 show that the spectrum of the partially loaded and partially dispersed system may be obtained by convolution of the spectra of the separate modulating waveforms about individual carriers.

Problems

3.1a A random variable of power 1 W has a power density spectrum $S(f)$ (symmetrical about d.c.) in the shape of an isosceles triangle and extends to a maximum frequency of 0·1 MHz. Draw the power density spectrum of the variable (normalised to unit total power)

(Ans. Triangle base 0·2 MHz, height 10 W/MHz)

b Draw the power density spectrum of a function formed by multiplying the variable of part a with a sinusoid of frequency 1 MHz and of unit total power.

(Ans. $0{\cdot}5\,S(f-1)+0{\cdot}5\,S(f+1)$)

c Draw the power density spectrum of a function formed by multiplying the variable of part b with a sinusoid of frequency 1·05 MHz and of unit total power.

Ans. $0{\cdot}25S(f-2{\cdot}05)+0{\cdot}25\,S(f+2{\cdot}05)+0{\cdot}25\,S(f-0{\cdot}05)+$ $0{\cdot}25\,S(f+0{\cdot}05)$. The low frequency terms add to give a trapezium of base 0·3 MHz, height 2·5 W/MHz)

NOTE. This problem illustrates (in an exaggerated way) the distortion obtained from frequency error in the local oscillator of a synchronous detector for amplitude modulation.

3.2. 240 channels of telephony each of 4 kHz bandwidth are assembled in f.d.m. The combined signal is clipped at 3 times its r.m.s. value and used to amplitude modulate (double sideband modulation) a 6 GHz carrier. At the receiver envelope detector, the total power of this wanted signal is 1 mW.

There is an interfering signal also at the demodulator input which has a flat topped spectrum (no residual carrier) of total bandwidth 20 kHz centred on the frequency 6000.1 MHz and of total power 1 μW.

Which telephone channels in the wanted system baseband have interference and what is their signal/interference noise ratio?

(Ans. $0{\cdot}1 \pm 0{\cdot}01$ MHz, 6·2 dB)

3.3 A lightly loaded 1 800 channel line-of-sight radio relay system (which can be taken as unmodulated for the purposes of this problem) has a test tone r.m.s. frequency deviation (f_d) of 140 kHz and has no pre-emphasis.

Interference occurs from a 300 channel line-of-sight radio relay system (using pre-emphasis) whose carrier is separated from that of the 1 800 channel system by 10 MHz.

Using Figure 3.5, which shows the r.f. spectrum of the 300 channel system, calculate the value of the interference transfer factor for a 4 kHz bandwidth channel situated at a frequency of 8 MHz in the baseband of the 1 800 channel system.

(Ans. 11 dB)

3.4 Derive expressions for the function $G_{u+w}(f)$ (equation 3.35 or 3.39) for the following interference situations:

Interference to a wide deviation frequency modulated system of mean square frequency deviation ΔF^2, modulated by a signal of Gaussian amplitude distribution. (Quasistationary approach applies.)

From a. An unmodulated carrier.

$$\left(\text{Ans. } G_{u+w}(f) = \frac{1}{\sqrt{(2\pi)}\Delta F}\, e^{-f^2/2\Delta F^2}\right)$$

b. A wide deviation frequency modulated system similar to the wanted system.

$$\left(\text{Ans. } G_{u+w}(f) = \frac{1}{2\sqrt{(\pi)}\Delta F}\, e^{-f^2/4\Delta F^2}\right)$$

c. A rectangular spectrum of total bandwidth B where the carrier is regarded as the centre frequency of the band.

$$\left(\text{Ans. } G_{u+w}(f) = \frac{1}{2B}\left[\text{erf}\,(a) - \frac{b}{|b|}\,\text{erf}\,(b)\right]\right.$$

$$a = (f + B/2)/\sqrt{2}\,\Delta F$$

$$\left. b = (f - B/2)/\sqrt{2}\,\Delta F\right)$$

Reference

1. RICHARDS, D. L., 'Statistical properties of speech signals', *Proc. I.E.E.* **111**, 941, (1964)
2. ABRAMSON, N., 'Bandwidth and spectra of phase-and-frequency-modulated waves', *I.E.E.E. Trans.*, **CS-11**, 407, (1963)
3. JOHNS, P. B., 'Interference between terrestrial line-of-sight radio-relay systems and communication satellite systems', *Electronics Letters*, **2**, 177, (1966)
4. FERRIS, C. C., 'Spectral Characteristics of f.d.m.-f.m. Signals', *I.E.E.E. Trans.*, **COM-16**, 233, (1968)
5. GLADWIN, A. S., 'Energy distribution in the spectrum of a frequency modulated wave', *Pt. I Phil. Mag.*, **35**, 787, (1944)
6. GLADWIN, A. S., 'Energy distribution in the spectrum of frequency modulated wave', *Pt. II Phil. Mag.*, **38**, 229, (1947)
7. SMITH, J. R. W. and SLOW, J. L., 'Energy distribution in a wave frequency modulated by a multichannel telephone signal', *A.T.E. J.*, **12**, 182, (1956)
8. MEDHURST, R. G., 'R. F. spectra of wave frequency modulated with white noise', *Proc. I.E.E.*, **107 C**, 314, (1960)
9. MEDHURST, R. G., HICKS, E. M., and GROSSETT, W., 'Distortion in frequency division multiplex f.m. systems due to an interfering carrier', *Proc. I.E.E.* **105 B**, 282, (1958)
10. BLACHMAN, N. M., 'Limiting frequency-modulation spectra', *Information and Control*, **1**, 26, (1957).
11. BLACHMAN, N. M., 'A generalization of Woodward's Theorem on FM spectra', *Information and Control*, **5**, 55, (1962)
12. MEDHURST, R. G., and ROBERTS, J. H., 'Evaluation of distortion in f.m. trunk radio systems by a Monte-Carlo method', *Proc. I.E.E.*, **113**, 570, (1966)
13. JOHNS, P. B., 'Graphical method for the determination of interference transfer factors between interfering frequency modulated multichannel telephony systems', *Electronics Letters*, **2**, 84, (1966).
14. JOHNS, P. B., 'Interference reduction factors in frequency modulated multichannel telephony systems', *I.E.E. Conference publication No. 39*, (1968)

15. MEDHURST, R. G. and ROBERTS, J. H., 'Expected interference levels due to interactions between line-of-sight radio relay systems and broadband satellite systems', *Proc. I.E.E.* **111,** 519, (1964)
16. MEDHURST, R. G., 'R.F. spectra and interfering carrier distortion in f.m. trunk radio systems with low modulation ratios', *IRE Trans.,* **CS-9,** 107, (1961)
17. JEFFERIS, A. K., 'A survey of interference problems associated with the use of digital satellite communications', *IEE Conference Publication No. 59,* (1969)

Chapter 4

ECHO AND AM TO PM DISTORTION IN FM SYSTEMS

4.1. INTRODUCTION

This chapter considers the distortion which arises in frequency modulation systems due to echoes. These may arise, for example, from a pair of mismatches separated by a length of transmission line. Thus if some energy is reflected by the mismatch farthest away from the energy source, and again at the mismatch closest to the source then some of the signal, after two reflections is again travelling in the same direction as the wanted signal and so appears at the load as an echo. Mode conversion-reconversion is another phenomena which gives rise to echoes in f.m. systems. This occurs in waveguide feeders at terrestrial radio-relay stations.

Some devices such as travelling wave tubes (t.w.t.) and transistor amplifiers exhibit a phenomenon whereby amplitude modulation at the input, caused by a number of modulated or unmodulated sinusoids (carriers) for example, appears at the output as phase modulation. In an f.m. system this will give rise to distortion of the wanted signal. Under certain conditions this distortion can be intelligible (Section 4.2.2). The estimation of this type of distortion is considered in this chapter.

4.2. ECHOES IN F.M. SYSTEMS

An echo in an f.m. system can be looked upon as a special case of phase modulation. Thus, following the argument used in Section 3.7, the sum of the wanted signal and unwanted interference is,

$$s(t) = A_w \cos [(\omega_c t+\phi_w(t)]+f_u(t) \cos [\omega_c(t+T)+\phi_w(t+T)] \quad (4.1)$$

where T is the time delay of the echo. Considering only the phase component of equation 4.1 i.e. ignoring the resultant amplitude variation caused by adding the signals,

$$s(t) = \cos [\omega_c(t)+\phi_w(t)+\mu(t)] \tag{4.2}$$

where $\mu(t)$ is the effective phase distortion, whose value is found by consideration of Figure 3.1. to be

$$\mu(t) = \frac{f_u(t) \sin [\omega_c T-\phi_w(t)+\phi_w(t+T)]}{A_w+f_u(t) \cos [\omega_c T-\phi_w(t)+\phi_w(t+T)]} \tag{4.3}$$

and, as before, for $f_u(t)$ small compared to A_w, equation 4.3 becomes

$$\mu(t) = v \sin [\omega_c T-\phi_w(t)+\phi_w(t+T)] \tag{4.4}$$

where

$$v = \frac{f_u(t)}{A_w}$$

The demodulated signal is $[\phi_w(t)+\mu(t)]$ so that the power spectrum of $\mu(t)$ is required. Writing

$$V(t) = \phi_w(t+T)-\phi_w(t) \tag{4.5}$$

he autocorrelation function of $\mu(t)$, $R_\mu(t)$, can be found by writing nto (1.16)

$$f(t) = \mu(t) = v \sin (\omega_c T+V(t)) \tag{4.6}$$

Recalling that $\omega_c T$ is a constant, and that $\sin \{\phi(t)\}$ and $\cos \{\phi(t)\}$ are uncorrelated, see Appendix 4, simple trigonometrical manipulation identical to Appendix 3, yields

$$R_\mu(\tau) = v^2\{R_s(\tau) \cos^2 \omega_c T+R_c(\tau) \sin^2 \omega_c T\} \tag{4.7}$$

where $R_s(\tau)$ indicates the autocorrelation function of $\sin (V(t))$
and $R_c(\tau)$ indicates the autocorrelation function of $\cos (V(t))$.
$R_s(\tau)$ and $R_c(\tau)$ were determined in Section 3.5.1, equation 3.11 et seq., so that using equation 3.13

$$\begin{aligned} R(\tau) &= v^2 e^{-R_v(0)} [\sinh R_v(\tau) . \cos^2 \omega_c T+\cosh R_v(\tau) \sin^2 \omega_c T] \\ &= v^2 e^{-R_v(0)} [e^{-R_v(\tau)}-e^{-R_v(\tau)} \cos 2 \omega_c T] \end{aligned} \tag{4.8}$$

Taking the Fourier transform of the autocorrelation function 4.8 gives the power spectrum as

$$W_\mu(f) = v^2[S_v(f) * \cos^2 \omega_c T . \delta(f)+C_v(f) * \sin^2 \omega_c T . \delta(f)] \tag{4.9}$$

where

$S_v(f)$ is the power spectrum of sin $(V(t))$
$C_v(f)$ is the power spectrum of cos $(V(t))$
$\delta(f)$ is the Dirac delta function discussed in Section 1.4.1.

Equation 4.9 is not in itself important. The technique used to obtain $W_\mu(f)$ is to expand equation 4.8 term by term and transform according to the procedure in equations 3.13 to 3.17. Thus,

$$S_v(f) = W_v(f) + \frac{W_v(f) \overset{3}{*} W_v(f)}{3!} + \ldots \tag{4.10}$$

and

$$C_v(f) = \delta(f) + \frac{W_v(f) \overset{2}{*} W_v(f)}{2!} + \ldots \tag{4.11}$$

where $W_v(f)$ is the power spectrum of $V(t)$
From (4.9)

$$\text{for} \quad \omega_c T = 2n\pi \pm \pi/2 \qquad W_\mu(f) = v^2\, S_v(f) \tag{4.12}$$

$$\text{for} \quad \omega_c T = n\pi \qquad W_\mu(f) = v^2\, S_v(f) \tag{4.13}$$

For intermediate values of $\omega_c T$ the spectrum is made up of the sum of these two spectra weighted by appropriate factors. Equations 4.10 and 4.11 show that solutions of this type of interference problem can be obtained from the method of successive convolution for systems with low mean square phase deviation. In the case of a rectangular modulating spectrum, for particular line-of-sight radio relay systems curves of echo amplitude (r) against time delay (T) are given in Reference 1.

4.3. INTERMODULATION DISTORTION FROM A.M. TO P.M. CONVERSION

Some devices in modern communications, e.g. travelling wave tubes (t.w.t.) and transistor amplifiers, have a phase modulation on the output signal which is dependent on the square of the envelope of the input signal, i.e. amplitude modulation is converted to phase modulation. Figure 4.1 shows a typical characteristic for a t.w.t.

Thus for all power levels, up to saturation, for example, the added phase modulation, $\theta(t)$ can be represented as a polynomial

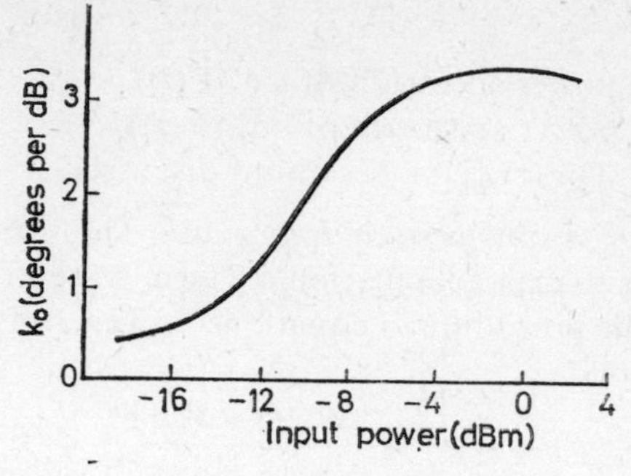

Figure 4.1 Typical variation of a.m. to p.m. conversion factor with input power for a travelling wave tube

of the envelope squared of the input signal, so that

$$\theta(t) = b_1(E_{\text{in}}^2(t)) + b_2\,(E_{\text{in}}^2(t))^2 + b_3(E_{\text{in}}^2(t))^3 \tag{4.14}$$

Taking as an example a number of signal carriers at different frequencies being applied to the t.w.t., in a satellite, for instance, a.m. to p.m. will cause distortion only if either

1. the difference in frequency between any two carriers equals the frequency difference between any two other carriers, or between either of the first-mentioned two carriers and a third. In this case the distortion is unintelligible, or
2. a device with gain/frequency slope precedes the device exhibiting a.m. to p.m. conversion. In this case intelligible crosstalk results.

The first, unintelligible, distortion results simply because the envelope amplitude of the two carriers beats at half the difference frequency and the envelope power beats at the difference frequency. When a constant times the difference frequency is added to the phase term of a third carrier, the combination frequency may be that of one of the carriers of the system.

The second, intelligible (crosstalk) distortion is very important, as it is more disturbing to a listener than unintelligible noise.

4.3.1. Unintelligible Noise from a.m. to p.m. Conversion

Let the input $V_{\text{in}}(t)$ consist of N frequency modulated carriers, thus,

$$\begin{aligned} V_{\text{in}}(t) &= \sum_1^N A_n \cos\,(\omega_n t + \phi_n(t)) = \sum_1^N A_n \cos\,(W_n(t)) \\ &= E_{\text{in}}(t).\cos\omega_0(t) \end{aligned} \tag{4.15}$$

where $E_{in}(t)$ is the envelope of $V_{in}(t)$ and is the parameter which determines the added phase modulation (equation 4.14). Consideration of the vector diagram of equation 4.15 referred to the arbitrary frequency ω_0, assuming the modulation small compared to the input power, gives

$$E_{in}(t) = \left[\sum_1^N (A_n)^2\right]^{1/2} \left[1 + \frac{\sum_{r=1}^N \sum_{s=1}^N A_r A_s \cos\,[W_r(t) - W_s(t)]}{\sum_1^N (A_n)^2}\right]^{\frac{1}{2}} \quad (4.16)$$

$(r \neq s)$ where only the two terms shown will be used.

The first part of equation 4.16 is a constant and the second is the amplitude modulation component. Using only the first term of equation 4.14, b_1 expressed in terms of a factor k_0 measured in degrees/dB is given for small k_0 by

$$b_1 = \frac{20 \log (1 + k_0)}{57 \cdot 3} = \frac{8 \cdot 69\, k_0}{57 \cdot 3} = 0 \cdot 1516\, k_0 \quad (4.17)$$

Thus, the output, with an added modulation $b_1 . [E_{in}(t)]^2$ to each carrier is

$$\sum_{n=1}^N A_n \cos \left[W_n(t) + 0 \cdot 1516\, k_0 \frac{\sum_{r=1}^N \sum_{s=1}^N A_r A_s \cos\,[W_r(t) - W_s(t)]}{\sum_{n=1}^N (A_n)^2} \right] \quad (4.18)$$

$$r \neq s$$

The original signal can be extracted from equation 4.18, by simple trigonometry, and remembering that k_0 is small, to give, for each of the N carriers in equation 4.18,

$$A_n \cos\,(W_n(t)) - A_n 0 \cdot 1516\, k_0 \left(\frac{\sum_{r=1}^N \sum_{s=1}^N \frac{A_r A_s}{2} \sin\,[W_n(t) \pm (W_r(t) - W_s(t))]}{\sum_{n=1}^N (A_n)^2} \right) \quad (4.19)$$

$$r \neq s$$

where the symbol $\pm$ indicates that both alternatives are present. The second term in equation 4.19 is the distortion term. Only if $W_n \pm (W_r - W_s)$ is equal to the angular frequency of another carrier will the distortion have any effect on a wanted signal. The worst case will occur for equally spaced carriers, cos $(W_n t)$, in the frequency domain. 4.19 is the most general equation, but in the special case of $A_n = A_r = A_s = A$, for N carriers 4.19 reduces to

$$A \sum_{n=1}^{N} \sum_{r=1}^{N} \sum_{s=1}^{N} \left(\cos W_n(t) - \frac{0{\cdot}1516\, k_0}{2N} \sin \left([W_n \pm (W_r - W_s)] t \right) \right)$$

$$r \neq s \qquad (4.20)$$

After some manipulation, as each combination of r and s in equation 4.20 appears twice, sin $(W_n \pm (W_r - W_s))$ is written as 2 sin $(W_n(t) + W_r(t) - W_s(t))$ then the carrier to intermodulation noise for one 'product' of the $r \neq s \neq n$ type in equation 4.20 is given by

$$\left(\frac{P_c}{P_I} \right)_N = \frac{N^2}{(0{\cdot}1516\, k_0)^2} \qquad (4.21)$$

Study of equation 4.20 shows that no intermodulation falls about the third harmonic frequency of the input, equation 4.15. There are also 'products' in equation 4.20 of the type ($n = r$ or $= s$). There are far fewer of these than ($r \neq s \neq n$) type and as it turns out that the intermodulation power of each is a quarter that of equation 4.21 they will be ignored. Table 4.1 gives the distribution of the ($r \neq s \neq n$) products for small numbers of carriers. It is easy to program a computer to remove the tedium involved with larger numbers of carriers.

Table 4.1 DISTRIBUTION OF ($r \neq s \neq n$) PRODUCTS

No. of Products Falling on Carrier No.	1	2	3	4	5	6	7	8
No. of Carriers								
1	0							
2	0	0						
3	0	1	0					
4	1	2	2	1				
5	2	4	4	4	2			
6	4	6	7	7	6	4		
7	6	9	10	11	10	9	6	
8	9	12	14	15	15	14	12	9

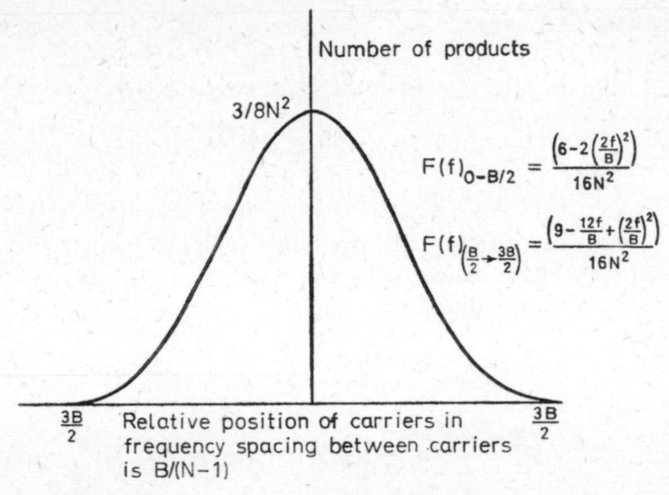

Figure 4.2 Distribution of numbers of intermodulation products for third order distortion for a large number of equally spaced carriers, N

For large numbers of carriers, (N), the number of products on the middle carrier is $3/8\ N^2$ and this number falls off to zero at $1\frac{1}{2}$ bandwidths away from the centre as the curve in Figure 4.2. Thus the carrier to total intermodulation noise power falling on the central carrier of a large number of equally spaced carriers will be

$$\left(\frac{P_c}{P_I}\right)_{n\to\infty} = \frac{10^3}{8{\cdot}64\ k_0^2} \tag{4.22}$$

Figure 4.1 shows a typical a.m. to p.m. coefficient (k_0) variation with input, and the maximum value of k_0 is about 3·4 degrees/dB. This corresponds to a carrier to intermodulation noise from a.m. to p.m. above of 10 dB.

4.3.2. Intel ligible Noise from a.m. to p.m. Conversion

Basically, an amplitude gain slope with frequency will give an amplitude modulation at the baseband rate on each carrier. This will then be added as an extra detectable phase modulation term by a.m. to p.m. conversion. In an f.m. system this causes intelligible crosstalk. Let the gain/frequency characteristic be given by

$$G = 1+g(\omega_o-\omega_n) \tag{4.23}$$

Let the input be given by equation 4.15, and this is then modified by equation 4.23 to

$$V_{\text{in}}(t) = \sum_{n=1}^{N} A_n G(\phi_n'(t)).\cos(\omega_n t + \phi_n(t)) \tag{4.24}$$

where
$$\phi_n'(t) = \frac{d(\phi_n(t))}{dt}$$

The analysis of Section 4.2.1 now holds where $A_n.G(\phi_n'(t))$ is substituted for A_n in equations 4.16 and 4.18. The amplitude modulation term, corresponding to equation 4.18 is

$$cA_n G(\phi_n'(t)).\cos\left\{W_n t + \frac{0{\cdot}1516\,k_0}{P_{ti}}\left[\sum_{q=1}^{N}(A_q)^2 \cdot g(\phi_q'(t)) + \sum_{r=1}^{N}\sum_{s=r+1}^{N} A_r A_s G(\phi_r'(t)).G(\phi_s'(t)).\cos(W_r t - W_s t)\right]\right\} \tag{4.25}$$

where
$$(G(\phi_n'(t)))^2 \approx 1 + 2.g(\phi_n'(t)) \text{ for small } g(\phi_n'(t))$$

$\frac{1}{2}\,\Sigma(A_n^2)$ is the total input power, which for convenience will be written as P_{ti}

c is the compression of the carrier power at the output

Note that the angular part of equation 4.25 has an r.f. component. Following the procedure used to obtain equation 4.19 from equation 4.18 gives the output due to the nth carrier for unequally spaced carriers as

$$cA_n\, G(\phi_n'(t)).\cos\left(W_n t + \frac{0{\cdot}1516\,k_0}{P_{ti}}\sum_{q=1}^{N} A_q^2 \cdot g(\phi_q'(t))\right)$$
$$- \frac{cA_n G(\phi_n'(t)).0{\cdot}1516\,k_0}{2P_{ti}}\sum_{r=1}^{N}\sum_{s=r+1}^{N} G(\phi_r'(t)).G(\phi_s'(t))$$
$$\times A_r A_s \sin\left[W_n t \pm (W_r t - W_s t) + \frac{0{\cdot}1516\,k_0}{P_{ti}}\sum_{q=1}^{N} A_q^2 \cdot g(\phi_q'(t))\right] \tag{4.26}$$

Summarising the mechanism of intelligible crosstalk, equation 4.26 is explained as follows.

Frequency modulation from each of the N carriers is converted to gain modulation by the gain/frequency curve of equation 4.23, and this gain modulation in turn is converted to intelligible phase information by a.m. to p.m. conversion. This appears as the linear term in the added phase $g(\phi_q'(t))$ in equation 4.26, and the significance of the two terms of equation 4.26 is, in order:

1. The modulation on each of the other carriers which pass through a gain slope followed by a.m. to p.m. conversion, is superimposed coherently on the baseband of any particular carrier as intelligible crosstalk, (even if the particular carrier does not have to pass through a gain slope itself prior to the t.w.t.).

2. If any particular carrier passes through a gain slope, and then into a multiple access t.w.t. exhibiting a.m. to p.m. conversion the modulation on the particular carrier will appear intelligibly in the baseband of each of the other carriers being amplified by the tube, together with the modulations of each of the other carriers.

Consider only the qth carrier to have passed through the gain slope characteristic. The crosstalk from the qth onto the nth carrier is made up as follows, using a certain amount of trigonometrical manipulation.

$$c_n A_n \cos\left[W_n t + \frac{0{\cdot}1516k_0}{P_{ti}} A_q^2 g(\phi_q'(t))\right]$$

$$-\sum_{s=1}^{N} c_s A_s^2 A_n \frac{0{\cdot}1516\,k_0}{P_{ti}} \sin\left[W_s t + \frac{0{\cdot}1516\,k_0}{P_{ti}} A_q^2 . g(\phi_q'(t))\right]$$

$$+(W_n t - W_s t)] - c_q A_q^2 (G\phi_q'(t))^2 \frac{0{\cdot}1516\,k_0}{P_{ti}} A_n$$

$$\sin\left(W_q t + \frac{0{\cdot}1516\,k_0}{P_{ti}} A_q^2 g(\phi_q'(t)) + (W_n t - W_s t))\right) \qquad s \neq q$$

where c_q is the compression of the qth carrier. The usual trigonometrical relationships allow this to be written as

$$Z_0 \cos\left(W_n t + \frac{0{\cdot}1516\,k_0}{P_{ti}} A_q^2 g(\phi_q'(t)) + \theta(t)\right)$$

where

$$Z_0 = A_n\left(c_n^2 + \left[\sum_{s=1}^{N} c_s A_s^2 \frac{0{\cdot}1516\,k_0}{2P_{ti}} + c_q A_q^2 (G\phi_q(t))^2 \frac{0{\cdot}1516\,k_0}{2P_{ti}}\right]^2\right)^{1/2}$$

$$\theta(t) = \tan^{-1}\left[\frac{\dfrac{0{\cdot}1516\,k_0}{2P_{ti}}\left\{\sum_{s=1}^{N} c_s A_s + c_q A_q^2 (G(\phi_q'(t))^2\right\}}{2c_n}\right] \qquad s \neq n \neq q \tag{4.27}$$

The time dependent terms that appear as added phase modulation in equation 4.27 are therefore given by

$$\frac{0{\cdot}1516\,k_0}{P_{ti}}\,.\,A_q^2\left[(g(\phi_p'(t))+\frac{c_q\,G(\phi_q'(t))^2}{2c_n}\right] \tag{4.28}$$

It must be realised that this is an approximate answer, as small argument approximations have been used throughout, and would appear to be justified by the small values for k_0 usually found in practice. Making the further approximation, used before, that

$$G(\phi_q'(t))^2 = 1+2g(\phi_q'(t))$$

and that for a linear gain slope

$$G(\phi_q'(t)) = g_0\phi_q'(t) \text{ (g_0 is a constant)}$$

the time dependent part (modulation) of equation 4.28 becomes

$$\frac{0{\cdot}1516\,k_0}{P_{ti}}\,A_q^2 g_0\phi_q'(t)\left(1+\frac{c_q}{c_n}\right) \tag{4.29}$$

If the amplitudes of the qth and nth carriers are the same, or if a large number of unequal amplitude carriers are being transmitted so that suppression of one by the other is negligible, then $c_q/c_n = 1$, so that the added modulation is

$$\frac{0{\cdot}1516\,k_0}{P_{ti}}\,.\,A_q^2\,.\,g_0\phi_q'(t) = g_0 0{\cdot}3032\,k_0\,\frac{2A_q^2}{A_1^2+A_2^2\,\ldots\,A_n^2\phi_q'(t)} \tag{4.30}$$

This expression is notable as it is independent of the number of carriers, depending only on the total power into the device.

For two equal amplitude carriers, one amplitude modulated by $g_0\phi_q(t)$

$$\text{Added} \qquad \phi(t)_2 = \frac{2\times 0{\cdot}1516\,k_0 A_q^2 g_0\phi_q'(t)}{P_{ti}} \tag{4.31}$$

From the considerations of Chapter 5, the range of c_q/c_n is from $\frac{2}{1}$ to $\frac{1}{2}$, these being the maximum suppressions of one carrier by another. When the carrier passing through the gain slope (the modulated carrier at the input) is very small compared to the unmodulated carrier at the input of the nonlinear device, the added phase modulation has a limit

$$\text{Added} \qquad \phi(t)_{2q} = 1{\cdot}5\,\frac{0{\cdot}1516\,k_0 A_q^2 g_0\phi_q'(t)}{P_{ti}} \tag{4.32}$$

The other condition with two carriers, that of a large qth (modulated) carrier, yields for the added phase modulation,

$$\text{Added} \qquad \phi(t)_{2q} = 3\,\frac{0{\cdot}1516\; k_0 A_q^2 g_0 \phi_q'(t)}{P_{ti}} \tag{4.33}$$

Finally, to a first approximation, for an infinite number of carriers, (where suppression of one carrier by another will be small) the formula is given by equation 4.32 where a difference in P_{ti} in the two cases must be noted.

These formulae hold for the addition of any baseband to any carrier being transmitted by the device.

Knowing k_0, $\phi_q(t)$ and the relative level of the carriers at the device input, for a given level of crosstalk (added phase modulation), the allowable gain slope can be specified.

References

1. MEDHURST R, G. 'Echo Distortion in Frequency Modulation', *Electronic and Radio Engineer*, **36**, 253–259, July 1959.

Chapter 5

NONLINEAR DISTORTION

5.1. INTRODUCTION

This chapter concerns itself with finding the output signal from a device when its nonlinear transfer characteristic (e.g. I_c versus V_b for a transistor) and input signal are known. A nonlinear characteristic is something of a two edged weapon. On the one hand none of the current wideband communication systems could work without it, while on the other, such characteristics can be the limiting factor in determining information capacity. For example, wideband systems which require transmission on a carrier frequency rely on modulators and frequency convertors yet amplifier saturation limits power output and causes distortion of the input signal which may render the information it carries undetectable at the output.

The output from any nonlinear device may be considered to be comprised of two components. These are the waveform identical to the input waveform and the waveform made up by subtracting this identical waveform from the output waveform at each instant in time. Notice that the first component of the output need not necessarily be of the same amplitude as the input waveform, and, indeed, in the case of an amplifier as the device, it should be many times the input amplitude. For the purposes of this chapter the second component will be referred to as the distortion, though in the case of a multiplier, part of this distortion would be selected as the wanted signal. The system designer often needs to know the quantity and nature of this distortion relative to the wanted signal at the output of such a device.

This chapter determines the quantity and nature of the distortion. However, the problem of nonlinear distortion has one constraint above that of most problems. Not only has the designer to apply a technique to a given problem, but he has to decide whether his problem can be solved by the currently available techniques or whether he must reformulate the problem as an approximation to his original problem so that a solution may be obtained. This is because, although any transfer characteristic which is continuous can be used and any input signal can be applied, all combinations of characteristic and signal do not have solutions. In the case of an inadmissible combination of signal and characteristic an approximation to one or other will often lead to good first order approximations to the practical answer. A second limitation to the answers available is that all phase information about the output is lost, as the solutions are given in terms of the power of the output at each frequency. This is as opposed to, for example, giving solutions in terms of the current, voltage and phase angle between them for each frequency of the output. Usually, however, this information is not required. The remainder of this chapter is divided into examination of the following three classes of nonlinearity.

5.1.1. Class I. General Nonlinear Devices

In this class, the output amplitude of the device is represented by a power series of the input amplitude (equations 5.1 and 5.2). In this approach any input/output transfer characteristic can be represented accurately, however, this approach can be used for an input consisting of a number of modulated or unmodulated carriers only if the power level of each carrier is the same. This class of nonlinearity is dealt with in Section 5.2.

5.1.2. Class II. Devices Represented by Odd Function Characteristics

In this class, input/output characteristics are symmetrical in the first and third quadrants. Thus if an input voltage V_{in} gives an output voltage V_{out}, then an input voltage $-V_{in}$ will give an output voltage $-V_{out}$ (see Figure 5.1). These are referred to as 'odd' function characteristics as the output voltage can be represented by a series of odd powers only of the input voltage. In this approach a suitable analytic function must be chosen as a compromise between simplicity of analysis and accuracy of the nonlinearity simulation. At the expense of inaccuracy in the representation of the transfer characteristic the range of input signal is extended over that in the previous example to include multiple carriers of differing levels, and signals with noise.

Whenever the signal and noise at the output is required only over the input frequency band this class of characteristic can be used, as only odd components of the transfer characteristic contribute to

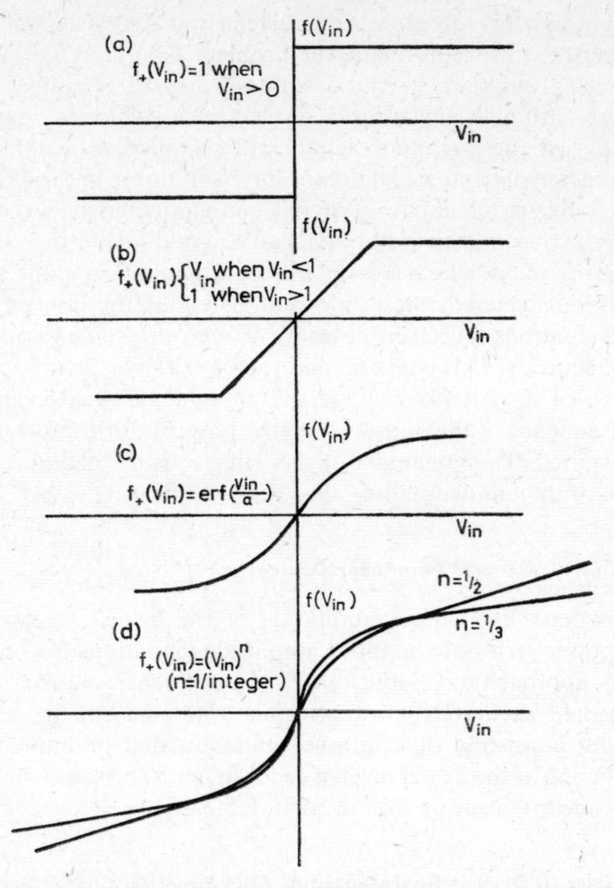

Figure 5.1 Classes of limiters analysed in this chapter. (a) The hard limiter; (b) The abrupt limiter; (c) The error function limiter; (d) The nth law limiter

in-band noise as will be shown in Section 5.2. This class of characteristic is considered in Section 5.3.

5.1.3. Class III. Devices Represented by Asymmetrical Characteristics

These characteristics describe such devices as rectifiers and multipliers and typical characteristics are shown in Figure 5.2. Usually devices such as these utilise their nonlinear characteristics to alter the signal frequency of interest. Thus noise considerations tend to play a minor role in the calculations. Asymmetrical characteristics are therefore included to complement the characteristics of the previous section. An identical technique to that used in the anal-

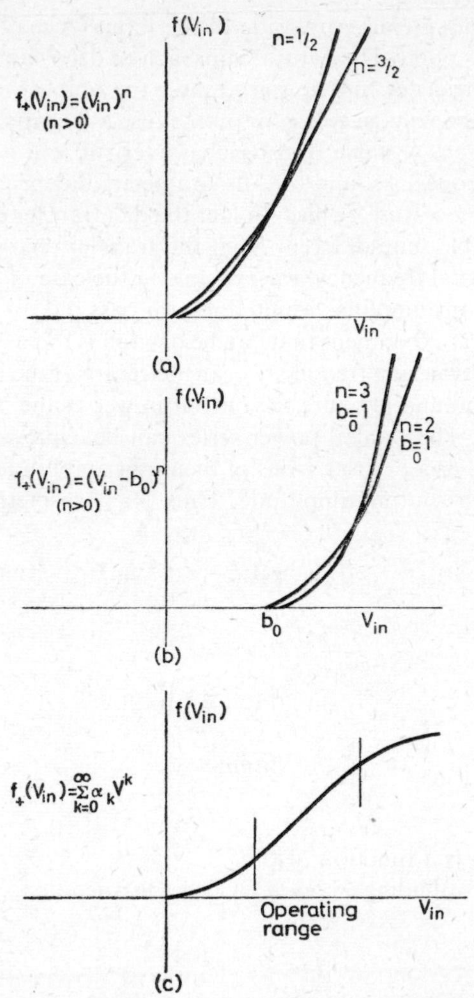

Figure 5.2 Classes of rectifier analysed in this chapter. (a) the nth law rectifier; (b) The biased nth law rectifier; (c) The small signal detector

ysis of odd function characteristics is fully applicable here and useful results are given in Section 5.4.

5.2. CLASS I. GENERAL NONLINEAR DEVICES

5.2.1. Representation of the Nonlinear Transfer Characteristic

In this section the coefficients of series representations of the general nonlinear characteristic are established. The criterion for the correct transfer characteristic is that it gives the correct output amplitude for every instantaneous amplitude of the input. A major difficulty appears at this point. The natural approach to determination of the input/output transfer function is to alter the voltage at the input and measure the voltage at the output i.e. use a d.c. approach. This is usually wrong. A single transfer characteristic will not be accurate for all frequencies, and for all parameters independant of the signal, e.g. temperature or bias. In fact the d.c. transfer characteristic will often be quite different from the transfer characteristic at the wanted signal frequency, for example in the case of a travelling wave tube or any amplifier which does not pass d.c.

Thus the best experiment that can be devised is to pass a sinusoid through the device at a frequency near the centre of the input signal frequency band and measure the output power as the input power is increased[12]. The wanted power series can be expressed as either a power series or a Fourier series of the input amplitude as a representation of the output amplitude. Thus the characteristic is written as:

$$V_{out} = \alpha_1 V_{in} + \alpha_2 V_{in}^2 + \alpha_3 V_{in}^3 + \ldots \qquad (5.1)$$

$$= \sum_{n=1}^{\infty} \alpha_n V_{in}^n$$

$$V_{out} = \beta_1 \sin \Omega V_{in} + \beta_2 \sin 2\Omega V_{in} + \ldots$$

$$= \sum_{m=1}^{\infty} \beta_m \sin m\Omega V_{in} \qquad (5.2)$$

Note that V_{in} is a function of time.
The problem of finding α_1, α_2, α_3 will be pursued first.

Let
$$V_{in} = V \sin(\omega t)$$

$$= V\left(\frac{e^{j\omega t} + e^{-j\omega t}}{2}\right) \qquad (5.3)$$

Thus the second term of equation 5.1 becomes

$$\alpha_2 V_{ia}^2 = \alpha_2 (V \sin \omega t)^2 = \alpha_2 V^2 \left(\frac{e^{j\omega t} + e^{-j\omega t}}{2}\right)^2$$

$$= \alpha_2 V^2 \left(\frac{e^{j2\omega t} + 2 + e^{-j2\omega t}}{2^2}\right)$$

$$= \alpha_2 V^2 \left(\frac{\sin 2\omega t + 1}{2}\right) \qquad (5.4)$$

Similarly the third term of equation 5.1 becomes

$$\alpha_3 V_{\text{in}}^3 = \alpha_3 (V \sin \omega t)^3 = \alpha_3 V^3 \left(\frac{e^{j\omega t} + e^{-j\omega t}}{2} \right)^3$$

$$= \alpha_3 V^3 \left(\frac{e^{j3\omega t} + 3e^{j\omega t} + 3^{-j\omega t} + e^{-j3\omega t}}{2^3} \right)$$

$$= \frac{\alpha_3 V^3}{2^2} (\sin 3\omega t + 3 \sin \omega t) \qquad (5.5)$$

Following this argument it is seen that

1. only the odd powers in the transfer characteristic equation 5.1 produce an output component at the input frequency.

2. only the central terms of the expansion of each of these odd powers contributes to the fundamental output (i.e. the same frequency as the input).

Thus the coefficient of sin (ωt) contributed by the third term of equation 5.1 is $3\alpha_3 V^3/2^2$, the coefficient contributed by the fifth term is $10\alpha_5 V^5/2^4$. In general for n odd

$$\alpha_n V_{\text{in}}^n = \alpha_n V_{\text{in}}^n \sin^n (\omega t)$$

$$= \alpha_n V^n \{k_n \sin (n\omega t) + k_{n-2} \sin (n-2)\omega t$$

$$\ldots + k_m \sin (m\omega t) \ \ldots \ k_1 \sin (\omega t)\} \qquad (5.6)$$

where $$k_m = \frac{1}{2^{n-1}} {}^nC_{\frac{n+m}{2}}$$

$$= \frac{n(n-1)(n-2)\ldots 1}{2^{n-1}\left[\left(\frac{n+m}{2}\right)\left(\frac{n+m}{2}-1\right)\ldots 1\right]\left[\left(n-\frac{n+m}{2}\right)\left(n-\frac{n+m}{2}-1\right)\ldots 1\right]}$$

Thus the general coefficient of the nth power (for n odd) of the power transfer characteristic is

$$\alpha_n \frac{V^n}{2^{n-1}} {}^nC_{\frac{n+1}{2}} \qquad (5.6a)$$

Thus, to recapitulate, if a sinusoidal test voltage $V \sin \omega t$ is applied to a device whose instantaneous transfer characteristic is

$$V_{\text{out}} = \alpha_1 V_{\text{in}} + \alpha_2 V_{\text{in}}^2 + \alpha_3 V_{\text{in}}^3 + \alpha_4 V_{\text{in}}^4 \ldots$$

then the output at frequency ω is

$$V_{out(\omega)} = \alpha_1 V \sin \omega t + \frac{\alpha_3 V^3}{2^2}\, {}^3C_2 \sin \omega t + \frac{\alpha_5 V^5}{2^4}\, {}^5C_3 \sin \omega t$$

$$= \left\{ \alpha_1 V + \frac{3}{4}\alpha_3 V^3 + \frac{10}{16}\alpha_5 V^5 + \frac{35}{64}\alpha_7 V^7 \ldots \right\} \sin \omega t \qquad (5.7)$$

Thus if the output voltage at this single frequency is plotted against the input voltage for positive and negative voltages the coefficients of the power series representing the resulting curve can be obtained.

This is achieved by standard curve fitting techniques or, more simply and accurately, by one of the standard computer programmes found in most libraries, which fits a polynomial to a set of Cartesian ordinates. If the polynomial representing the test result is written as

$$V_{out(\omega)} = \alpha_{t1} V + \alpha_{t3} V^3 + \alpha_{t5} V^5 \ldots \qquad (5.8)$$

then each coefficient α_{tk} is related to the corresponding coefficient α_k of the wanted transfer characteristic in a unique way. Thus the α_k can be found.

Example 5.1

The transfer characteristic of an amplifier is tested by varying the power input of a 300 MHz signal. The output power is measured by a monitor whose response is 20 dB lower at 600 MHz than at 300 MHz, i.e. it may be considered to respond to the fundamental only.

If the measured characteristic is as shown in Figure 5.3 what instantaneous amplitude transfer characteristic should be used?

Solution

The first step is to convert this characteristic to an amplitude characteristic. If the input impedance of the amplifier is known then this power characteristic of Figure 5.3 can be converted to a voltage characteristic. Otherwise (as here) the impedance should be assumed to be 1 Ω and the output will be in the in the same units. As Figure 5.3 has a logarithmic scale, and the polynomial analysis is of a linear curve, accuracy will be improved if the points whose coefficients are fed into the analysis are closer together as the output power level increases. The equation of the amplitude

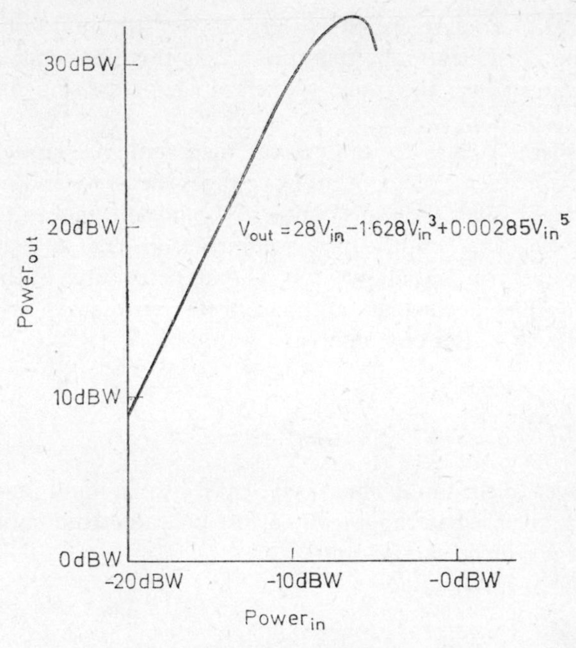

Figure 5.3 Power transfer characteristic of an amplifier. Note; a normalised impedance at input and output of 1 Ω is used

characteristic representing the curve of Figure 5.3 is found by computer to be

$$V_{out} = 28{\cdot}0\,V_{in} - 1{\cdot}628V_{in}^{3} + 0{\cdot}00285V_{in}^{5}$$

This equation corresponds to equation 5.8, and by using the equivalence between equations 5.8 and 5.7 the wanted characteristic is

$$V_{out} = 28{\cdot}0\,V_{in} - 2{\cdot}171\,V_{in}^{3} + 0{\cdot}00456V_{in}$$

If the transfer characteristic is represented by equation 5.3 then the first job is to find the value of β_1, β_2, etc. Again it is assumed that the device has been tested with a sinusoidal input and the voltage output at the input frequency is measured as the input voltage is varied. The approach is to find the instantaneous voltage input/output characteristic as a power series by the procedure described in the previous paragraphs. Once this characteristic has been established, then the coefficient β_1, β_2, etc., can be determined by Fourier analysis of its relevant range. Ideally, the input range to be covered extends from the peak of the input voltage waveform in the positive direction to that in the negative direction. However, the input may be noise-like in its waveform, so that occasionally a peak will be many times the r.m.s. voltage of the waveform. It is impractical to analyse the characteristic over all input voltages to obtain the coefficients of equation 5.2 as this characteristic can only be known over the range of the voltage of the sinusoidal test signal.

The general rule is to analyse the instantaneous input/output characteristic over a range of input voltages (negative and positive) equal to eight times the r.m.s. value of the input signal to the non-linear device. This empirical rule comes from the statistics of a random waveform and allows only a small percentage of the input to be transferred through an inaccurate representation of the input/output transfer characteristic.

Example 5.2

If the power of the input signal is 0·5 mW into an input impedance of 100 Ω, over what range of the test characteristic should the Fourier analysis be carried out?

Solution

A 0·5 mW signal into 100 Ω, corresponds to an r.m.s. voltage of V volts given by

$$\frac{V^2}{100} = 0{\cdot}5 . 10^{-3}$$

Therefore $$V = 0{\cdot}222 \text{ V r.m.s.}$$

Therefore the range of input voltage should be least to ±0·888 V.

5.2.2. The Monte Carlo Method

A very powerful approach to any nonlinear problem is the Monte Carlo method. This usually involves the availability of a very large computer. The input signal is represented by a very large number of sinusoids of random phase. This input is processed through the transfer characteristic (equations 5.1 or 5.2) and the phase and amplitude of the constituent sinusoidal components of the result of the passage of the input is obtained by Fourier analysis. These output sinusoids are then summated, those in phase with the input being recorded as the wanted signal, and the rest being added together as the noise. This method was discussed in connection with frequency modulation (see Section 3.5.6) which is, of course, the passage of a baseband through a nonlinear device with an exponentially shaped transfer characteristic.

Obviously the degree of computation necessary depends on the complexity of the input waveform. All cases could be handled analytically in that if time and patience were available one could produce the correct answers as only simple trigonometrical relationships between powers of sinusoids and multiple angles are required. However even the substitution of an input

$$V_{\text{in}} = A\cos\omega_1 t + B\cos\omega_2 t + C\cos\omega_3 t$$

into

$$V_{\text{out}} = \alpha_1 V_{\text{in}} + \alpha_2 V_{\text{in}}^2 + \alpha_3 V_{\text{in}}^3 + \ldots + \alpha_r V_{\text{in}}^r + \ldots$$

can clearly be seen to be tedious.

When n sinusoids are fed into a device whose nonlinear characteristic is represented by an odd order power series, a formula for the amplitude of each component of the output is available[1].

$$\alpha_r . \frac{r!}{2^{r-1}} \frac{\{A^a B^b \ldots\}\cos\{(a-2a')\omega_1 \pm (b-2b')\omega_2 \pm \ldots\}t}{\{(a-a')!\,(b-b')!\ldots\}\{a'!b'!\ldots\}} \tag{5.9}$$

Where each of the quantities a, b, etc. is zero or positive so that $a+b+\ldots = r$ and each of the quantities a', b', etc. is zero or a positive integer, with the restriction that

$$a' \leqslant \tfrac{1}{2}a, \qquad b' \leqslant \tfrac{1}{2}b \text{ etc.}$$

However, most inputs that correspond to practical situations are composed of continuous spectra, i.e. a large number of sinusoids would be needed to synthesise them. Thus the solution of equation 5.9 when a practical input is being considered is clearly too cumbersome to be performed normally and a computer will be necessary. In fact, in using the Monte Carlo method with practical input spectra, long computer times are often necessary. Thus an approach that deals with continuous spectra inputs would seem to be worth considering. Such an approach is considered in the next section.

5.2.3. Signal/Noise Ratio for an Input of Random Noise (Power Law Device)

If a random noise input, of limited bandwidth is fed into a nonlinear device whose characteristic is represented by equation 5.1, then the output consists of a component which is identical to the input, and incoherent components. It is often required to know the ratio of wanted to unwanted signal in a given spectral region.

The key to the analysis of this section (and of the next two) is given in the next sentence. If the spectrum of the output is required, the best way to obtain this is to form the autocorrelation function of the output, as the spectrum and autocorrelation function are related by the Fourier transform (equation 1.19). The importance of this relationship cannot be underestimated. The autocorrelation function is defined by

$$R(\tau) = \lim_{T\to\infty} \frac{1}{2T} \int_{-T}^{+T} f(t).f(t+\tau)\, dt \tag{1.16}$$

If we now write the output from the nonlinear device as $f(t)$ then the usefulness of the method will depend on the ability to rearrange the resultant equations and to establish their physical meaning. Thus

$$f(t) = V_{out}(t) = \alpha_1 V_{in}(t) + \alpha_2 V_{in}^2(t) + \alpha_3 V_{in}^3(t) \ldots$$

so that (1.16) can be written as

$$R_{out}(\tau) = \lim_{T\to\infty} \frac{1}{2T} \int_{-T}^{+T} \left(\alpha_1 V_{in}(t) + \alpha_2 V_{in}^2(t) + \alpha_3 V_{in}^3(t)\right)\left(\alpha_1 V_{in}(t+\tau)\right.$$
$$\left. + \alpha_2 V_{in}^2(t+\tau) + \alpha_3 V_{in}^3(t+\tau) \ldots \right) dt$$
$$= \lim_{T\to\infty} \frac{1}{2T} \int_{-T}^{+T} \{\alpha_1^2 V_{in}(t).V_{in}(t+\tau)\}\, dt$$
$$+ \lim_{T\to\infty} \frac{1}{2T} \int_{-T}^{+T} \{\alpha_2^2 V_{in}^2(t).V_{in}^2(t+\tau)\}\, dt \ldots$$

This equation can also be written (see equations 1.15 and 1.16) as

$$R_{out}(\tau) = E(\alpha_1^2 V_{in}(t).V_{in}(t+\tau)) + E(\alpha_2^2 V_{in}^2(t).V_{in}^2(t+\tau))$$
$$+ E(\alpha_1\alpha_2 V_{in}(t)\, V_{in}^2(t+\tau)) + \ldots \tag{5.10}$$

Each of the terms in equation 5.10 is a multiple moment (cf. equation 1.12). Half the terms in equation 5.10 are odd moments, so that by equation 1.39a these become equal to zero. The even moments will be treated by equation 1.39b to enable these to be broken down into the manageable second moments $E[V_{in}(t).V_{in}(t+\tau)]$ the autocorrelation function of the input, and $E[(V_{in}^2(t)]$ or $E[(V_{in}^2(t+\tau)]$ which represents the power of the input as it is the mean square value of the input waveform. The procedure will be indicated by examining the first few terms of equation 5.10, in the light of

Example 1.3, and assuming $V_{in}(t)$ is Gaussian

$$E(\alpha_1^2 V_{in}(t).V_{in}(t+\tau)) = \alpha_1^2 R_{in}(\tau)$$
$$E(\alpha_1\alpha_3 V_{in}(t).V_{in}^3(t+\tau)) = 3\alpha_1\alpha_3 R_{in}(\tau) P_{in}$$
$$E(\alpha_3\alpha_1 V_{in}(t+\tau).V_{in}^3(t)) = 3\alpha_3\alpha_1 R_{in}(\tau) P_{in}$$

If the device can be represented by equation 5.1 with powers one and three only, one more term needs to be added to those above. This was already calculated in Example 1.3 as

$$E(\alpha_3^2 V_{in}^3(t) V_{in}^3(t+\tau)) = \{9R_{in}(\tau).P_{in}^2+6R_{in}^3(\nu)\}\alpha_3^2$$

Thus equation 5.10 becomes for a cubic transfer characteristic,

$$R_{out}(\tau) = R_{in}(\tau)(\alpha_1^2+6\alpha_1\alpha_3 P_{in}+9\alpha_3 P_{in}^2)+6\alpha_3^2 R_{in}^3(\tau) \quad (5.11)$$

Before proceeding to generalise equation 5.11 to a higher power of transfer characteristics equation 5.11 will be transformed into the frequency domain by use of equation 1.19. Using the fact that multiplication transforms into convolution (see Appendix I) and denoting the convolution of two spectra $S_1(f)$ and $S_2(f)$ by

$$S_1(f) * S_2(f) \quad \text{then}$$

$$S_{out}(f) = S_{in}(f)(\alpha_1+3\alpha_3 P_{in})^2+6\alpha_3^2 S_{in}(f) * S_{in}(f) * S_{in}(f) \quad (5.12)$$

Equation 5.12 corresponds in nonlinearity theory to $1+1 = 2$ in basic arithmetic, in that it contains the basic ingredients to allow progress to all more complicated transfer characteristics. Examining equation 5.12 term by term, the first term corresponds to the input spectrum itself, and is the component of the output which is identical to the input waveform and is therefore (usually) the wanted signal. It is made up of an amplified component, and a component due to the cubic law throwing up part of the output in phase with the fundamental. The second term is the incoherent part of the cubic term resultant and is often referred to as third order intermodulation. This has a spectral shape given by the convolution of the input spectra with itself twice. Convolution has been discussed in detail in Chapter 3 and the reader is referred back to Section 1.3.2 for more detail. A term such as $R_{in}^2(\tau)$, due to the square law term in equation 5.1, has a spectrum given by the convolution of the input spectrum with itself. If a narrow band input is assumed (less than an octave bandwidth in the case of the square law term) then no part of this spectrum falls back into the input signal band as noise. This contribution is either about zero frequency or the

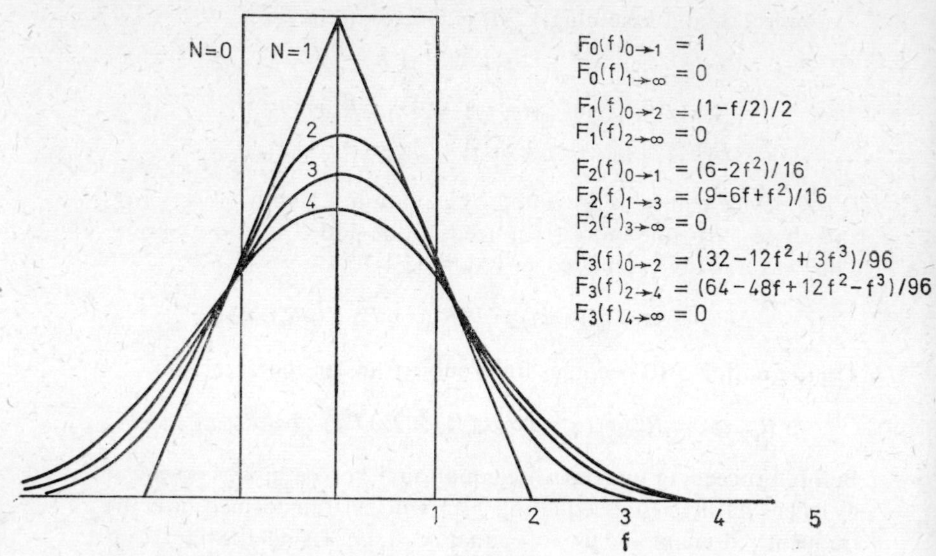

Figure 5.4 The spectral shape of the convolution of a rectangle with itself n times. Note: The area of each shape (i.e. power) is unity

second harmonic. This merely means that if the transfer characteristic has even power in its representation the output will not have a zero mean level.

It can be shown that, in general, the even power can be neglected as far as the signal/noise ratio in the input signal band is concerned, as long as the signal band is narrow. For this reason in the rest of this section the transfer characteristic will be considered to be of odd powers only.

The fifth power of equation 5.1 will be expected to contribute to the third order intermodulation from a term

$$E(\alpha_5^2 V_{\text{in}}^5(t) \, . \, V_{\text{in}}^5(t+\tau)) \quad \text{in equation 5.10}$$

The number of combinations of second moments in this tenth joint moment is ${}^5C_2 . {}^5C_2 . (5-2)!$ to give terms in $R_{\text{in}}^3(\tau)$ i.e. choose two out of five $V_{\text{in}}(t)$ and two out of 5 $V_{\text{in}}(t+\tau)$ to give R_{in}^2, then only the combinations of the remainder concern us. The other terms in which α_3 is included, automatically allow the third order intermodulation to be written as

$$6\alpha_3^2 \left(1 + \frac{\alpha_5}{\alpha_3} 10 . P_{\text{in}}\right)^2$$

An extension of this principle leads to general formulae applicable to calculation of the spectral density of the output signal and each component of noise (e.g. third order or fifth order intermodulation) thus the component of the output in phase with the input is given by

$$S_{out}(f) = S_{in}(f)\left\{\alpha_1 + 3\alpha_3 P_{in} + 15\alpha_5 P_{in}^2 + 105\alpha_7 P_{in}^3 + \ldots \right.$$

$$\left. \frac{(2r+1)!}{2^r r!}.\alpha_{2r+1}.P_{in}^r\right\}^2 \tag{5.13a}$$

The third order distortion has a spectral density given by

$$6\left\{S_{in}(f) \overset{3}{*} S_{in}(f)\right\}\left\{\alpha_3 + 10\alpha_5 P_{in} + 105\alpha_7 P_{in}^3 \ldots \right.$$

$$\left. \frac{(2r+3)!}{3!2^r.r!}\alpha_{2r+3} P_{in}^r\right\}^2 \tag{5.13b}$$

The symbol $\overset{3}{*}$ indicates the convolution of the spectrum with itself and with itself again as before. The kth order intermodulation is given by (k odd)

$$S_{in}(f) \overset{k}{*} S_{in}(f).k! \quad \left\{\sum_{\substack{n=k \\ n \text{ odd}}}^{\infty} \alpha_n P_{in}^{\left(\frac{n-k}{2}\right)} {}^nC_k.(n-k-1)\right.$$

$$\left.(n-k-3)\ldots 5.3.1\right\}^2 \tag{5.14}$$

One fairly common form of input spectrum is the rectangular spectrum. In this case the multiple convolution of the rectangular input spectrum gives the spectral shape shown in Figure 5.4.

However, and this must be noted carefully, not all of the power of the spectrum $S_{in}(f) \overset{3}{*} S_{in}(f)$, nor of higher order spectra, falls about the input frequency band of $S_{in}(f)$. This is seen to be true by considering the argument of Section 5.2.1, where in equation 5.5 for example, it can be seen that $\frac{3}{4}$ of the amplitude spectrum of $\sin^3(\omega t)$ falls back onto $\sin(\omega t)$. A similar argument for the power spectrum of $S_{in}(f)$ leads to the same distribution of the powers across the frequency band as the distribution of amplitudes in the expansion of $\sin^n(\omega t)$ where n is an integer.

If, therefore, it is required to find the noise power from the nth order noise, for example, where n is odd, then

1. Find the coefficients of the nonlinear transfer characteristic from Section 5.2.1.
2. Write out the general nth order noise equation 5.14.

3. The proportion of the power of $S_{in}(f) \overset{n}{*} S_{in}(f)$ that falls back in band is given by equation 5.6a, i.e. the proportion is

$$\frac{1}{2^{n-1}} \cdot {}^{n}C_{\frac{n+1}{2}}$$

4. Its spectral shape is taken from the convolution of $S_{in}(f)$ with itself $n-1$ times. Call this shape, normalised to unit power $S_n(f)$. The proportion of this falling into the wanted frequency band f_1 to f_2 is given by

$$\int_{f_1}^{f_2} S_n(f).df$$

5. Finally combine these to give the total power in the band f_1 to f_2 by multiplying the proportion of the power of $S_{in}(f) \overset{n}{*} S_{in}(f)$ that falls into the fundamental band by the fraction of the spectrum $S_n(f)$ that falls across the wanted frequency band, and this then replaces $S_{in}(f) \overset{n}{*} S_{in}(f)$ in the equation of (2) above, with the coefficients of (1) inserted.

It is now possible to vary the input power P_{in} and study the variation in noise power. This technique can easily be extended to cover all the components of noise of interest.

Example 5.3

The input to the amplifier of Example 5.1 is an amplitude modulated carrier, with the modulation signal being multiplex telephony extending from zero frequency to 100 kHz with constant power.

What is the signal/noise ratio at the centre of the output band of the amplifier?

Solution

Taking an input amplitude of unity for the input signal, the corresponding spectral density at the centre of the third order distortion is $\frac{3}{4}$ (From Figure 5.4.) while that for fifth order distortion is $\frac{230}{384}$. These numbers refer to the spectral density at midband relative to a unity power 'flat' input spectrum.

Thus the carrier level at the output of the amplifier is from equation 5.13 and the result of Example 5.1,

$$P_c = P_{in}[28-3(2{\cdot}171)P_{in}+(0{\cdot}00456)15.P_{in}^2] \qquad (5.15a)$$

For a transfer characteristic polynomial of fifth power only third and fifth order distortions occur. Using the final two equations of this section, the power of these distributions is given by

$$P_{d3} = \tfrac{3}{4}\,6.P_{\text{in}}^{3}[-2{\cdot}171+10(0{\cdot}00456)P_{\text{in}}]^{2} \qquad (5.15\text{b})$$

$$P_{d5} = \tfrac{230}{384}\,5.4.3.2.1.(0{\cdot}00456)^{2}P_{\text{in}}^{5} \qquad (5.15\text{c})$$

Equation 5.13 is only valid over a certain range of input power. Over this range of input power the signal/noise ratio can be shown by equation 5.15 to be as in Figure 5.5.

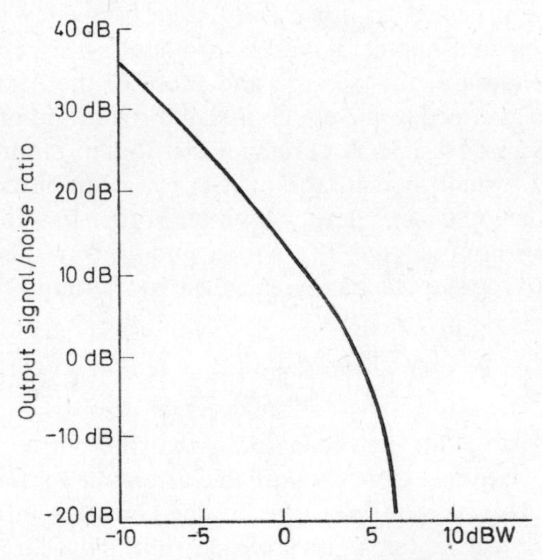

Figure 5.5 Output signal/noise ratio from the amplifier of Figure 5.3

With enough computational power available, accuracy of signal/noise ratio at any frequency can be determined by this approach for any nonlinearity. Its limitation is that all inputs are treated as being noiselike. This means in practice that suppression of one signal by another due to this simultaneous passage through the nonlinearity cannot be computed.

5.2.4. Signal/Noise Ratio for an Input of Random Noise (Sum of Sinusoids Representation)

The output voltage of a nonlinear device can be formulated in terms of a series of powers of the input voltage (equation 5.1) or a series of harmonically related sinusoidal functions of the input (equation 5.2).

This section deals with the latter representation, where

$$V_{out} = \sum_{m=1}^{\infty} \beta_m \sin m\Omega V_{in}$$

The main advantage of this method over that of Section 5.2.3 is that, in general, less terms are required to give a certain percentage accuracy of representation of nonlinearity, while beyond the range of input voltage over which precise representation is required the 'sum of sinusoids' approach departs more slowly from the true curve. Furthermore, the approach is elegant, it closely resembles one of the approaches to noise analysis in frequency modulation systems given in Chapter 3, and it provides an excellent bridge between the ideas of this section and those of the next. Random noise will be assumed to have a Gaussian distribution of amplitude levels (see Section 1.3.3). A Gaussian distributed input waveform spends only a small percentage of its time at amplitudes greater than four times the r.m.s. level. A function that has an amplification A at low input voltage, V_{in}, with a smooth transition to a constant output voltage Ak has been given by Lanning and Battin[2] as

$$V_{out} = \frac{Ak}{32}\left[38 \sin\left(\frac{V_{in}}{2k}\right) + 7 \sin\left(\frac{3V_{in}}{2k}\right) + \sin\left(\frac{5V_{in}}{2k}\right)\right] \tag{5.16}$$

The advantage of this characteristic is that the output remains at the constant output voltage Ak until the instantaneous input voltage reaches $4k$. Thus if the r.m.s. level of the Gaussian input remains below k this will represent an excellent limiting characteristic to which the subsequent method can be applied.

Specific devices can be simulated by the procedure given in Section 5.2.1.

Once again, the key to the analysis is to find the autocorrelation function as a first step to finding signal/noise ratio in the output, and the spectra of the signal and noise components. The autocorrelation function and the spectral density form a Fourier transform pair. The autocorrelation function is

$$R_{out}(\tau) = \lim_{T\to\infty} \frac{1}{2T} \int_{-T}^{+T} f(t).f(t+\tau)\, dt \tag{1.16}$$

$$= \lim_{T\to\infty} \frac{1}{2T} \int_{-T}^{+T} \left(\sum_{k=1}^{\infty} \beta_k \sin k\Omega V_{in}(t)\right) \left(\sum_{l=1}^{\infty} \beta_l \sin l\Omega V_{in}(t+\tau)\right) dt \tag{5.17}$$

As in the previous Section (5.2.3) the 'expectations' notation, E, first introduced in Section 1.3.5 will be used. Thus the autocorrelation function 5.17 is written as

$$R_{out}(\tau) = \sum_{k=1}^{\infty} \sum_{l=1}^{\infty} \beta_k \beta_l E(\sin k\Omega V_{in}(t) . \sin l\Omega V_{in}(t+\tau)) \quad (5.18)$$

The subtlety in this section comes in recognising that the right hand side of this equation can be split by use of

$$\text{Sin } \theta = \frac{e^{j\theta} + e^{-j\theta}}{2j} \quad (5.19)$$

into a sum of joint probability density functions of the variables $V_{in}(t)$ and $V_{in}(t+\tau)$ (see Section 1.3.7 especially equation 1.28). This can then be related to combinations of autocorrelation functions of the input, so allowing identification of 'signal' and 'noise' components. Using equation 5.19 the typical term of the r.h.s. of equation 5.18 becomes

$$E[\sin \{k\Omega V_{in}(t)\} . \sin \{l\Omega V_{in}(t+\tau)\}] = E\{\sin (V_1 x_1) \sin (V_2 x_2)\}$$

$$= E\left\{\frac{e^{jV_1 x_1} - e^{-jV_1 x_1}}{2j} \cdot \frac{e^{jV_2 x_2} - e^{-jV_2 x_2}}{2j}\right\}$$

$$= \tfrac{1}{4} E\{e^{j(V_1 x_1 - V_2 x_2)} + e^{j(V_2 x_2 - V_1 x_1)} - e^{j(V_1 x_1 + V_2 x_2)} - e^{-j(V_1 x_1 + V_2 x_2)}\} \quad (5.20)$$

The expectation of exp $[j(V_1 x_1 + V_2 x_2)]$, which is the same as the joint characteristic function of the two variables, is related to the expectation of two variables (the autocorrelation function of the input) by equation 1.35

Thus equation 5.20 becomes

$$E[\sin \{k\Omega V_{in}(t)\} \sin \{l\Omega V_{in}(t+\tau)\}]$$

$$= \tfrac{1}{2} \{e^{R_{in}(\tau) kl\Omega^2} - e^{R_{in}(\tau) kl\Omega^2}\} e^{-R_{in}(0) \cdot \Omega^2 \left(\frac{k^2 + l^2}{2}\right)} \quad (5.21)$$

Thus summing all similar terms, with the knowledge that

$$\sinh (z) = \frac{e^z - e^{-z}}{2}$$

(5.18) becomes

$$R_{out}(\tau) = \sum_{k=1}^{\infty} \sum_{l=1}^{\infty} \beta_k \beta_l \, e^{-R_{in}(0)\Omega^2 \left(\frac{k^2 + l^2}{2}\right)} \sinh (\Omega^2 kl R_{in}(\tau)) \quad (5.22)$$

where $R_{out}(\tau)$ is the autocorrelation function of the output
$R_{in}(0)$ is the input power and is independent of frequency
$R_{in}(\tau)$ is the autocorrelation function of the input.

Taking the Fourier transform 1.17 of equation 5.22 term by term necessitates the examination of

$$\int_{-\infty}^{\infty} \{\sinh (K.R_{in}(\tau))\}\, e^{-j\omega\tau}\, d\tau$$

where K is a constant

Using the expansion

$$\sinh z = z + \frac{z^3}{3!} + \frac{z^5}{5!} + \ldots$$

Normalising the autocorrelation function

$$\frac{R_{in}(\tau)}{R_{in}(0)} = \varrho(\tau) \quad \text{by definition}$$

and noting that the Fourier transform of $\varrho(\tau)$ is the spectrum of unit power $w(f)$ it can be seen that the Fourier transform of $\sinh (K.R_{in}(\tau))$ is

$$KR_{in}(0).w(f) + \frac{K^3(R_{in}(0))^3}{3!} w(f) \overset{3}{*} w(f)$$
$$+ \frac{K^5(R_{in}(0))^5}{5!} w(f) \overset{5}{*} w(f) \qquad (5.23)$$

where $R_{in}(0)$ is the total input power to the limiter.

$$K = \Omega^2 lk.$$

The first term of equation 5.23 is the input spectrum amplified by K. The second term is third order intermodulation and involves convolution of the input spectrum with the convolution of the input spectrum with itself. The third term is fifth order intermodulation.

A nonlinear device as described by equation 5.16 will give nine terms for equation 5.22 and each will have a portion coherent with the input (first term of equation 5.23) and third order intermodulation corresponding to the second term of equation 5.23. Thus, for example, taking the first three terms of the generalised series of equation 5.2, the first term of equation 5.22 is

$$R_{out}(\tau)_{fund} = R_{in}(\tau)\Omega^2\{\beta_1^2\, e^{-\Omega^2 R_{in}(0)} + 4\beta_2^2\, e^{-\Omega^2 4R_{in}(0)}$$
$$+ 9\beta_3^2\, e^{-\Omega^2 9R_{in}(0)} + 4\beta_1\beta_2\, e^{-\Omega^2(5/2)R_{in}(0)} + 12\beta_2\beta_3\, e^{-\Omega^2(13/2)R_{in}(0)}$$
$$+ 6\beta_1\beta_3\, e^{-\Omega^2 5R_{in}(0)}\} \qquad (5.24)$$

Where $R_{in}(\tau)$ is the transform of the input spectrum. The input signal does not have to be a number of equal amplitude carriers, as long as the number is great enough so that the instantaneous amplitude distribution is approximately Gaussian.

Equation 5.24 shows that the input spectrum $w(f)$ is multiplied by a constant: i.e. the output spectrum is identical to the input spectrum. Thus the relative amplitudes of the constituent parts of the input cannot alter, so that if the input consists of a number of modulated carriers, none of them can take more than their share of output power.

With only a few carriers and a highly nonlinear device one would expect one carrier to emerge with more than its share of power.

The question arises as to how many carriers (or sinusoids) might reasonably be transferred through a nonlinear device with a maintenance of accuracy. Bennett[3] has given theoretical curves for the envelope distributions of six to one hundred amplitude sinusoidal waveforms. These indicate that ten carriers modulated or unmodulated, can be considered to have an approximately Gaussian distribution of amplitude. In the next section, a technique is given which allows calculation of the suppression of one carrier by another in a nonlinear device. It is shown in Figure 5.10, that in a practical amplifier no more than 0·5 dB suppression may be expected if four or more carriers are present, even when one carrier is very small compared to the rest of the carriers.

Example 5.4

A mainland-island radiotelephone link has four 4 kHz speech channels in each direction, with s.s.b.-s.c. working. The final power amplifier is to deliver 10 W to the aerial. The combination of aerial gains and channel separation filters results in a 20 dB discrimination against transmitter power entering the associated receiver. The path loss is 35 dB, and the carrier to intermodulation noise ratio is not to exceed 20 dB. There is effectively no frequency separation of 'go' and 'return' channels.

A power amplifier is available, having a characteristic given by equation 5.16, with a low-level gain of 40 dB, and limiting at an instantaneous 'voltage' of 10 V ($= Ak$ in equation 5.16).

Is this a suitable amplifier for this link?

Solution

Three questions must be answered before the specification can be met. Is the output power large enough? Is the carrier-to-intermodulation noise ratio in the wanted band high enough? Is the ratio of received carrier level to intermodulation noise power 'spread' into the received channels by the transmitter nonlinearity high enough?

The 'voltage' of equation 5.16 is the square root of the power in Watts, from the way in which it is used in equation 5.17. Equation 5.22 is the important equation in this problem. Here k and l take values 1, 3, and 5. As the low level gain is 40 dB, A in equation

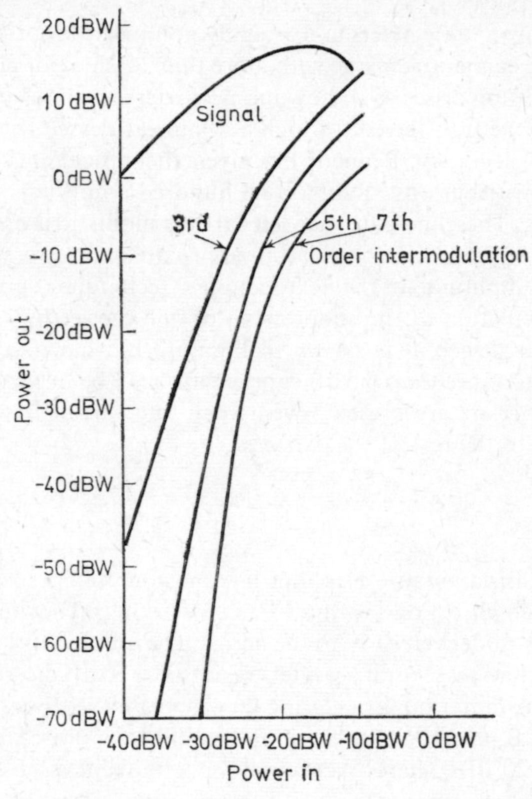

Figure 5.6 Output signal and intermodulation power variation with input (see Example 5.4)

5.16 is 100, and so $k = 0{\cdot}1$, as the limiting voltage is 10. $\Omega = 5 = \frac{1}{2}k$. The expansion of the sinh term in equation 5.22 gives the odd order intermodulation spectra in turn, term by term. A simple computer programme written to solve equation 5.22 with the above parameters and those of equation 5.16, gave the powers of the carrier, $R_{in}(0)$, and the third, fifth and seventh order intermodulation powers as shown in Figure 5.6. It can be seen that no further orders of intermodulation are necessary.

The first question posed above is therefore answered; the amplifier will deliver 10 W.

The next two questions require some thought about the relative strengths of the in-band and 'spread' intermodulation powers at the receiver input. The in-band noise will be the same strength relative to the transmitter power at the input to the receiver as it was at the output of the transmitter. These values can be read directly from Figure 5.6, where it is seen that the sum of the intermodulation powers does not violate the 20 dB signal/noise specification.

In the case of the noise spread into the receiver by the transmitter at the same site, where the transmitters and receivers at each site will be considered to be identical, the signal level at the receiver will be 35 dB below the output signal power at the transmitter, while any power in the wanted band at the output of the associated transmitter will be reduced by 20 dB, due to filtering. Thus relatively speaking, the signal/total intermodulation noise will be 15 dB worse than the values of Figure 5.6. However, only a certain proportion of this intermodulation noise will fall into the adjacent channels. The spectrum of the wanted signal is 16 kHz wide, and is rectangular. Thus, the intermodulation spectra are given by Figure 5.4, where it is seen that the spectral density for the third and fifth order spectra is about half that of the original spectrum at the edge of the channels. The energy falling into the 4 kHz band alongside the 16 kHz spectrum from the third order intermodulation is given, as a percentage of the total third order intermodulation power, by the technique discussed at the end of section 5.2.3

$$= \left(\int_{1}^{1^1/_2} (9 - 6x + x^2)/16 \, dx \right) 100\,\% \cdot \tfrac{3}{4}$$

$$= 7{\cdot}5\%$$

This channel is the most affected by the intermodulation power, so that it receives the power of the third order intermodulation as given by Figure 5.6 reduced by less than 12 dB. Taking the 15 dB effective carrier level reduction mentioned before together with about 12 dB reduction of total intermodulation power the signal/noise ratio is 3 dB worse relatively than the figures given by Figure 5.6 so that for third order distortion alone the power is 25 dB below the signal. The spectral shape of the higher order distortions need not be considered as below the +10 dBW point on the wanted signal power curve of Figure 5.6 the fifth and higher order intermodulation power curves are at least 15 dB below that of the third order intermodulation power curve, and thus may be neglected.

The amplifier is thus suitable for the link, though in this latter respect the margin for error is small.

Note that at low drive levels (− 40 dBW to − 35 dBW) the slopes of the curves in Figure 5.6 are in almost exact agreement with theory, in that if the order of the curve is N, the slope is N dB increase in output power for every 1 dB increase in drive power.

Another interesting feature of Figure 5.6 is that the gain is exactly A^2, i.e. 40 dB for low drive levels, as is to be expected. The onset of saturation is slow, as the output voltage of equation 5.16 does not fall off until relatively high instantaneous voltages are reached, as compared to the characteristic of Figure 5.3, which is not untypical of a travelling wave tube, wherein once the electrons in a bunch have passed each other, space charge forces rapidly accelerate the debunching effect which contributes to the loss in gain.

5.3. CLASS II. DEVICES REPRESENTED BY ODD FUNCTION CHARACTERISTICS

These functions are symmetrical in the first and third quadrants and can be represented by polynomials of odd powers only. Examples of such functions are shown in Figure 5.1.

The advantage of the method given here is that accurate computation of the output can be made, including suppression of one sinusoid by another (i.e. the relative levels of two sinusoids can differ between output and input). These sinusoids could represent unmodulated carriers in a radio communication system for example.

The disadvantage is that only certain representations of the input/output transfer characteristic can be used. However, the results for suppression from this section, coupled with the results of Section 5.2 should allow analysis of most characteristics and inputs.

The method used here is referred to as the *characteristic function method*. Before considering particular results the theory of the method will be outlined. The reader wishing to avoid the chore of understanding the principle of the method is advised to step directly to equation 5.38 as the derivation of this key equation is the purpose of the next two sections.

5.3.1. The Characteristic Function Method

As with the direct method of Section 5.2 the characteristic function method has as its object the autocorrelation function of the output of the nonlinear device. The autocorrelation function is the statistical average of the output at time (t) times the output at time ($t+\tau$) over all τ, (equations 1.15 and 1.16), and its Fourier transform is the power spectral density equations 1.19 and 1.20. In the direct method of Section 5.2 results were obtained by putting 'directly'

$$V_{out} = f(V_{in}) \tag{5.25}$$

into the equation defining the autocorrelation function 1.16. The right hand side was expanded to allow treatment as a series of multiple moments, in order to use the formula 1.38.

In the characteristic function method, inverse Fourier transform of the Fourier transform of the transfer characteristic of equation 5.25 is substituted into the autocorrelation function 1.16. This may appear to unnecessarily complicate the procedure, as taking the inverse Fourier transform of a Fourier transform of a function leads back to the function, but the reader may recall occasions when multiplying both sides of an equation by the same factor aided solution! In this case the beauty of the method lies in the fact that where equation 1.16 is written out fully, the constants of integration allow a regrouping of integrals so that equation 1.16 becomes the multiple of separate functions of input and of the device itself. The function of the input signal is the characteristic function, and this in turn produces the major advantage of the technique, because the characteristic function of the total input is the multiple of the separate characteristic functions of each statistically independent part of the input (e.g. separate modulated carriers and/or noise). From these separated integrals the output amplitude of each separated constituent part of the input can be obtained.

The steps indicated will now be set out in detail. Let equation 5.25 be the transfer characteristic. The Fourier Transform of this is called the transfer function and is given by

$$F(u) = \int_{-\infty}^{\infty} f(V_{in})\, e^{-juV_{in}}\, dV_{in} \tag{5.26}$$

Thus the output of the device can be expressed in terms of the input by the inverse Fourier transformation.

$$V_{out} = f(V_{in}) = \frac{1}{2\pi} \int_{-\infty}^{\infty} F(u) e^{jV_{in}u}\, du \tag{5.27}$$

For the purpose of writing down the autocorrelation function 1.16 V_{out} is to be multiplied by a replica of itself shifted by time τ. Let equation 5.27 correspond to time t, and let q be the transform variable corresponding to time $(t+\tau)$, then

$$R_{out}(\tau) = \underset{T\to\infty}{\text{Limit}} \frac{1}{2T} \int_{-T}^{+T} \frac{1}{2\pi} \int_{-\infty}^{\infty} F(u) e^{jV_{in}u} \, du \frac{1}{2\pi} \int_{-\infty}^{\infty} F(q) e^{jV_{in}(\tau)q} \, dq \, dt \tag{5.28}$$

But the characteristic function of V_{in} at (t) and at $(t+\tau)$ is, from equation 1.28

$$M(u, q, \tau) = \underset{T\to\infty}{\text{Limit}} \frac{1}{2T} \int_{-T}^{+T} e^{j(uV_{in}+qV_{in}(\tau))} \, dt \tag{5.29}$$

where it is recalled that V_{in} is, of course, a function of time, t. As equation 5.29 represents all the time dependent functions in equation 5.28, then equation 5.28 may be rewritten, neatly, as

$$R_{out}(\tau) = \frac{1}{4\pi^2} \int_{-\infty}^{\infty} F(u) \int_{-\infty}^{\infty} F(q) \, M(u, q, \tau) \, dq \, du \tag{5.30}$$

And the output spectrum is given by

$$W_{out}(f) = \int_{-\infty}^{\infty} R_{out}(\tau) \, e^{-j\omega\tau} \, d\tau \tag{5.31}$$

Notice that the equation for the transfer characteristic 5.28 is the continuous analogy to the equation for the transfer characteristic as a sum of sinusoids (equation 5.2). If this point is appreciated, it can be seen that this method, although appearing complicated at first sight, is related to the method of Section 5.2.4 in the same way as the Fourier transform is related to Fourier analysis.

The work that has been performed since the introduction of this method has been devoted to the solution of equation 5.30 for the inputs of one or more sinusoids of any amplitude with or without noise. The simplest such input to the simplest (analytically speaking) device will be the subject of Section 5.3.2. However, before proceeding it is necessary to introduce a mathematical nicety which is strictly necessary but is here introduced to allow the reader who is using this work as a primer to 'recognise' the equations of the literature.

If a device is represented by,

$$\begin{aligned} f_+(V_{in}) &= f(V_{in}) && \text{when} \quad V_{in} > 0 \\ &= 0 && \text{when} \quad V_{in} = 0 \\ f_-(V_{in}) &= -f(V_{in}) && \text{when} \quad V_{in} < 0 \end{aligned} \tag{5.32}$$

e.g. is any of the characteristics in Figure 5.1, then, due to the discontinuity at the origin and convergence difficulties, the Fourier transform must be replaced by the Laplace transform (for a more detailed consideration of this point see Reference 4).

The relevant Laplace transform is

$$F(p) = F_+(p)+F_-(p) = \int_{-\infty}^{\infty} f(V_{in})\, e^{-pV_{in}}\, dV_{in} \qquad (5.33)$$

Thus changing from the Fourier transform to the Laplace transform allows analysis of characteristics with sharp corners and those for which the output does not return to zero as the input level is increased.

5.3.2. General Formula for the Output Autocorrelation Function of Signal and Noise Input

This section extends the previous one to a specific input of any number of sinusoids (unmodulated carriers) and independent (Gaussian) noise.

Let the input be the sum of signal and noise, thus,

$$V_{in}(t) = s(t)+n(t) \qquad (5.34)$$

where $s(t)$ represents the signal and $n(t)$ represents the independant noise. The object is to put equation 5.30 into a more 'computable' form. This effectively means that the characteristic function must be found. As the signal and noise are independent, the characteristic function of the sum is the multiple of the two separate characteristic functions (see Section 1.3.7). Thus equation 5.30 becomes

$$R_{out}(\tau) = \frac{1}{(2\pi j)^2} \int F(p_1)\, dp_1 \int F(p_2)\, dp_2 M_s(p_1p_2\tau) M_n(p_1p_2\tau) \qquad (5.35$$

The integrals are contour integrals, $M_s(p_1p_2\tau)$ is the joint characteristic function of $s(t)$ and $s(t+\tau)$ and $M_n(p_1p_2\tau)$ is the joint characteristic function of $n(t)$ and $n(t+\tau)$. When the noise is random and of zero mean, then use can be made of the complex version of equation 1.35 which gives $M_n(p_1p_2\tau)$, thus equation 5.35 becomes

$$R_{out}(\tau) = \frac{1}{(2\pi j)^2} \int F(p_1)\, e^{\frac{R_n(0)\, p_1^2}{2}} \int F(p_2)\, e^{\frac{R_n(0) p_2^2}{2}}$$
$$e^{R_n(\tau) p_1 p_2}\, M_s(p_1p_2\tau)\, dp_1\, dp_2 \qquad (5.36)$$

The objective is now to split equation 5.36 into separate integrals to allow their evaluation. The 'signal' characteristic function cannot be touched yet as the exact nature of the signal has not yet been specified. However the exponential term can be expanded as a power series

$$e^{R_n(\tau)p_1p_2} = \sum_{k=0}^{\infty} \frac{R_n(\tau)^k p_1^k p_2^k}{k!} \tag{5.37}$$

so that equation 5.36 becomes

$$R_{\text{out}}(\tau) = \sum_{k=0}^{\infty} \frac{R_{\text{in}}(\tau)^k}{k!(2\pi j)^2} \int F(p_1)p_1^k \, e^{\frac{R_n(0)\,p_1^2}{2}} \int F(p_2)p_2^k$$
$$e^{\frac{R_n(0)p_2^2}{2}} \, M_s(p_1p_2\tau) \, dp_2 \, dp_1 \tag{5.38}$$

Equation 5.38 is the autocorrelation function of the output of a nonlinear device for a signal and noise input. It is an infinite series because a nonlinear characteristic would be expected to give rise to a multiple component output as in Section 5.1. It depends on the noise power at the input, $R_n(0)$, and the signal characteristic function. The next section considers one particular input, a sinusoid, with the noise.

5.3.3. An Input of One Sinusoid and Noise

In this section one sinusoid in noise will be considered to a general characteristic. In the next section multiple signals to a particular device will be considered.

Let $s(t)$ in equation 5.34 be given by

$$s(t) = A \cos(\omega_c t + \phi) = A \cos(\theta(t)) \tag{5.39}$$

The joint characteristic function for this signal is, from equation 1.28, (writing p for jv),

$$M_s(p_1p_2\tau) = E(e^{p_1A \cos \theta(t) + p_2A \cos \theta(t+\tau)}) \tag{5.40}$$

This can be expanded using

$$e^{z \cos p} = \sum_{m=0}^{\infty} e_m I_m(z) \cos mp \tag{5.41}$$

where $e_0 = 1$, $e_m = 2$ $(m > 0)$

and $I_m(z)$ is a tabulated constant dependant on m and z. $I_m(z)$ is given explicitly in Reference 5, page 187

p and t are independent in equation 5.40, p being associated with the input characteristic and t with the input signal, the 'E' of equation 5.40 can be split into the product of 'averagings' giving

$$M_s(p_1p_2\tau) = \sum_{m=0}^{\infty} \sum_{n=0}^{\infty} e_m e_n E(I_m(p_1A) . I_n(p_2A))$$
$$E(\cos m\theta(t) . \cos n\theta(t+\tau)) \tag{5.42}$$

The final 'averaging' of equation 5.42 can be simplified as it is only non-zero in the case of $m = n$. The reader can show for himself from the definition (equation 1.15) that

$$E(\cos m\theta . \cos n\theta) = 0 \quad \text{for} \quad m \neq n \tag{5.43}$$
$$= \frac{\cos m\theta}{2} \quad \text{for} \quad m = n \neq 0$$
$$= 1 \quad \text{for} \quad m = n = 0$$

Furthermore, in equation 5.42

$$E(I_m(p_1A).I_n(p_2A)) = I_m(p_1A).I_n(p_2A) \tag{5.44}$$

as the time average of a constant is the constant itself. Notice that if the amplitude of the sinusoid in equation 5.39, A, were not constant, i.e. if it were modulated by noise or signal, equation 5.44 would not hold, so that it is at this point that the fact that a discrete sinusoid is present is recognised. Combining equations 5.43, 5.42 and 5.38, with $n = m$, and manipulating the resultant equation, gives

$$R_{\text{out}}(\tau) = \sum_{k=0}^{\infty} \sum_{m=0}^{\infty} \frac{R_n(\tau)^k . e_m}{k!} \left\{ \frac{1}{(2\pi j)} \int F(p).p^k.I_m(pA)\, e^{\frac{R_n(0)p^2}{2}}\, dp \right\}^2 \times \cos(m\omega_c\tau) \tag{5.45}$$

In equation 5.45 the expression in the brackets is a constant and represents an amplitude and adopting the (now) traditional notation of Rice[6], equation 5.45 is written

$$R_{\text{out}}(\tau) = \sum_{k=0}^{\infty} \sum_{m=0}^{\infty} \frac{R_n(\tau)^k}{k!} e_m h_{mk}^2 \cos(m\omega_c\tau) \tag{5.46}$$

where

$$h_{mk} = \frac{1}{2\pi j} \int F(p).p^k.I_m(pA)\, e^{\frac{R_n(0)p^2}{2}}\, dp \tag{5.47}$$

An equation equivalent to equation 5.46 is often the starting point in the literature (Equation (1) of Cahn[7], Equation (4) of Jones[8], and Equation (1) of Shaft[9]). The physical significance of equation 5.46 is seen immediately upon its expansion, thus

$$R_{\text{out}}(\tau) = h_{00}^2 + 2\sum_{m=1}^{\infty} h_{m0}^2 \cos(m\omega_c\tau) + \sum_{k=1}^{\infty} \frac{h_{0k}^2}{k!} (R_n(\tau))^k + 2\sum_{m=1}^{\infty}\sum_{k=1}^{\infty} \frac{h_{mk}^2}{k!} (R_n(\tau))^k \cos(m\omega_c\tau) \tag{5.48}$$

In order, these terms represent,

(a) The constant part of the device output (i.e. the d.c. level).
(b) The periodic part of the output due to the interaction of the signal with itself, for example, the harmonics in the case of a single sinusoid.
(c) The noise part of the output due to interaction of noise with itself.
(d) The interaction of signal and noise.

The $(h_{mk})^2$ are the respective powers of these components on transforming (equation 5.46) so that h_{mk} is the amplitude at the output of each component. An example will now be devoted to the application of equation 5.46 to a nonlinear device.

5.3.4. Signal and Noise at the Output of a Hard Limiter

Example 5.5

How does the output signal power and output noise power depend on the input signal/noise ratio for a device whose amplitude characteristic is described by Figure 5.1a? Assume that only the signal over the input frequency band is of interest.

Solution

The output spectrum for any device is given by the inverse Fourier transform 5.48. Thus, using equation 1.19

$$\begin{aligned} W_{\text{out}}(f) &= \int_{-\infty}^{\infty} R_{\text{out}}(\tau)\mathrm{e}^{-\mathrm{j}\omega\tau}\,\mathrm{d}\tau \\ &= h_{00}\delta(f) + \sum_{m=1}^{\infty} h_{m0}^2(\delta(f+mf_c)+\delta(f-mf_c)) \\ &\quad + \sum_{k=1}^{\infty} \frac{h_{0k}^2}{k!}\left(W_n(f)\overset{k}{*}W_n(f)\right) \\ &\quad + \sum_{m=1}^{\infty}\sum_{k=1}^{\infty} \frac{h_{mk}^2}{k!}\left(W_n(f+mf_c)\overset{k}{*}W_n(f+mf_c)\right. \\ &\quad \left. + W_n(f-mf_c)\overset{k}{*}W_n(f-mf_c)\right) \end{aligned} \tag{5.49}$$

where $\overset{k}{*}$ indicates the kth convolution of the spectrum with itself (see Section 1.3.2, Equation 1.7 and Figure 1.3) and $\delta(f-f_1)$ is the delta impulse function at frequency $f = f_1$ (see Section 1.4.1). The h_{mk} are given by equation 5.47. $W_n(f)$ is the power spectral density of the input noise at frequency f. First, the signal power will be computed, as this involves computation of a single term. Then the noise power, which involves multiples terms of equation 5.48, will be discussed. As pointed out in the last section, the second term of equation 5.49 is the 'signal' term. It is a series of sinusoids at positive and negative harmonics of the input signal. Positive and negative frequencies are required for the purposes of the convolution process. The fundamental signal power is therefore the sum of the power at $\pm f_c$. Therefore,

$$\text{Signal Power} = 2h_{10}^2 \tag{5.50}$$

Thus equation 5.47 must be evaluated for $m = 1$, $k = 0$. In equation 5.47, $F(p)$ is the Laplace transform of the device characteristic (equation 5.33). The transfer function equation 5.33 therefore becomes, for $f(V_{\text{in}}) = 1$ in equation 5.32

$$F(p) = \int_{-\infty}^{\infty} 1\mathrm{e}^{-pV_{\text{in}}}\,\mathrm{d}V_{\text{in}} = \frac{1}{p} \tag{5.51}$$

Equation 5.47 is a contour integral, which is somewhat off-putting, but with a contour along the imaginary axis and in the positive real part of the complex plane (see Reference 10, page 302), equation 5.47 upon the substitution of equation 5.51, reduces to the more manageable

$$h_{mk} = \frac{(-1)^{\frac{k+m-1}{2}}}{\pi} \int_0^{\infty} V^{k-1} J_m(VA)\,\mathrm{e}^{-\frac{R_n(0)V^2}{2}}\,\mathrm{d}V \tag{5.52}$$

Where $J_m(z)$ is a close relation of $I_m(z)$ (See Appendix 5) and is also tabulated in Reference 5, page 144. $R_n(0)$ is the power of the noise at the input and $A^2/2$ is the signal power at the input. Equation 5.52 is now able to be computed numerically as the ratio of $A^2/2R_n(0)$, i.e. as the ratio of input signal/noise, varies.

Like a large number of other common integrals, that in equation 5.52 has been given a name by the mathematicians. In this case it is Hankels exponential and its solution involves the confluent hypergeometric function, so that the reader writing an algorithm for the amplitude h_{mk} in the case of hard limiter, may well wish to build it up from the following equation,

$$h_{mk} = \frac{\left(\frac{A^2}{2R_n(0)}\right)^{m/2} {}_1F_1\left(\frac{m+k}{2}; m+1; -\frac{A^2}{2R_n(0)}\right)}{\left(\frac{R_n(0)}{2}\right)^{k/2} \Gamma(m+1)\Gamma\left(1-\frac{m+k}{2}\right)} \quad (5.52a)$$

$(m+k)$ odd,

where $\Gamma(u)$ is the Gamma function whose properties are given in Appendix 5, where it is defined as

$$\Gamma(u) = \int_{-0}^{\infty} e^{-t} t^{u-1}\, dt$$

and $\quad (u)\Gamma(u) = \Gamma(u+1), \quad$ and $\quad \Gamma(\tfrac{1}{2}) = \sqrt{\pi}$

${}_1F_1$ = Confluent Hypergeometric Function, where

$${}_1F_1(a; c; z) = 1 + \frac{a.z}{c.1!} + \frac{a(a+1)z^2}{c(c+1)2!} + \ldots \quad (5.53)$$

and $\quad \dfrac{A^2}{2R_n(0)} = (S/N)_{in} = \dfrac{\text{Input Signal Power}}{\text{Input Noise Power}}.$

Two limiting cases of equation 5.53 of interest are as follows:

$$\text{As } (S/N)_{in} \to 0 \quad {}_1F_1(a; c; z) \to 1 \quad (5.54)$$

$$\text{As } (S/N)_{in} \to \infty \quad {}_1F_1(a; c; z) \to \frac{\Gamma(m+1)}{\Gamma\left(\frac{m+2-k}{2}\right) . (S/N)_{in}^{\frac{k+m}{2}}} \quad (5.55)$$

Thus from equations 5.53 and 5.50 and Appendix 5, the signal power out of a hard limiter (Figure 5.1a) of unity output amplitude is

$$S_{out} = \text{Output Signal Power} = 2h_{10}^2 = 2\left\{\frac{\sqrt{(S/N)_{in}}}{\Gamma(2)\Gamma(\frac{1}{2})}\right\}^2$$

$$\times\{{}_1F_1(\tfrac{1}{2}; 2; -(S/N)_{in}\}^2 = (2/\pi)(S/N)_{in}\{{}_1F_1(\tfrac{1}{2}; 2; -(S/N)_{in}\}^2 \quad (5.56)$$

The variation of signal power out against signal/noise ratio at the input has been computed from equation 5.56 and is shown in Figure 5.7.

The limiting cases for large and small signals relative to the noise power are obtained from equations 5.56, 5.54 and 5.55 (see problem 5.2). Thus,

$$\begin{aligned} &\text{As} \quad (S/N)_{in} \to 0, \quad S_{out} \to (2/\pi)\,(S/N)_{in} \\ &\text{As} \quad (S/N)_{in} \to \infty, \quad S_{out} \to (2/\pi).(4/\pi) \end{aligned} \quad (5.57)$$

Now the noise part of equation 5.48 must be considered. The solution follows exactly the same steps as for the signal, but, of course, in this case, many more terms must be included. Two 'types' of noise arise, namely those discussed as (c) and (d) after equation 5.48. The third term of equation 5.48 contributes noise falling back across the band due to intermodulation of noise with itself. Only this term would exist if the relative signal level tended to zero. It is assumed that the noise is narrow-band in order that the indicated convolutions of equation 5.49 do not 'spread' the noise spectrum so much that the tails of the spectra at the harmonics $\pm f_c$ become significant. Only the odd k give contributions to the fundamental as shown in Section 5.2.1.

The proportion of the kth convolution that falls into band was calculated in Section 5.2.1. (Equation 5.6a). Thus the coefficients of the noise output, not surprisingly, are identical to those of equation 5.7. The noise from these terms is, therefore,

$$N_{\text{out}(n\times n)} = \frac{h_{01}^2}{1!} R_n(0) + \frac{3h_{03}^2}{4.3!}(R_n(0))^3 + \frac{10h_{05}^2}{16.5!}(R_n(0))^5$$

$$\ldots \frac{h_{0k}^2 . R_n(0)^k}{2^{k-1}\left(\frac{k-1}{2}\right)!\left(\frac{k+1}{2}\right)!} \qquad k \text{ odd} \qquad (5.58)$$

The fourth term of equation 5.48 is the interaction of the signal and noise. The concept of successive convolutions can be extended to find the contribution of this term to the fundamental band. The point is that a signal harmonic and a noise harmonic 'beat' together to produce an in-band component. The question is, which harmonics, and how much falls back in band? The answer is that the total noise power is

$$N_{\text{out}(s\times n)} = \frac{1}{2}\cdot\frac{2h_{21}^2}{1!}R_n(0) + \left(\frac{3}{2^2}\cdot\frac{2h_{12}^2}{2!} + \frac{1}{2^2}.2\frac{h_{32}}{2!}\right)R_n^2(0)$$

$$+\left(\frac{4}{2^3}2.\frac{h_{23}^2}{3!} + \frac{1}{2^3}.2.\frac{4_{43}^2}{3!}\right)R_n^3(0)$$

$$+\left(\frac{10}{2^4}.2\frac{h_{14}^2}{4!} + \frac{5}{2^4}.2.\frac{h_{34}^2}{4!} + \frac{1}{2^4}.2.\frac{h_{54}^2}{4!}\right)R_n^4(0)$$

$$+\ldots \qquad (5.59)$$

The coefficient of $R_n^k(0)$ is a series made up of

$$\sum_{n=1}^{n\leqslant(k+1)/2} \frac{2.h^2_{(k+3-2n).k}}{k!}\cdot\frac{P_n}{2^k}$$

P_n is the nth term from the edge of the $(k+2)$th line in Pascals Triangle, which is

1	$k = 1$
1 1	$k = 2$
1 2 1	$k = 3$
1 3 3 1	$k = 4$
1 4 6 4 1	$k = 5$
1 5 10 10 5 1	$k = 6$
1 6 15 20 15 6 1	$k = 7$

etc.

Thus the general term is

$$\frac{(k+1)!\, h_{mk}^2 . R_n(0)^k}{\left(\frac{k+1+m}{2}\right)!\left(\frac{k+1-m}{2}\right)!\, 2^{k-1} k!}$$

$$\text{all } k \quad m < k+2 \quad k+3-2n = m \text{ integer} \tag{5.60}$$

It is now possible to compute the sum of equation 5.58 and 5.59. It may appear that as the number of terms increases, the significance of each becomes greater due to the $R_n^k(0)$ term. In fact, h_{mk} has a term which completely cancels this out (see equation 5.53), so that a close study reveals that direct feed through is the most significant contribution. Figure 5.7 shows output noise variation with input signal/noise ratio. Note that all values have been 'normalised' by choosing a characteristic with a maximum output level of unity.

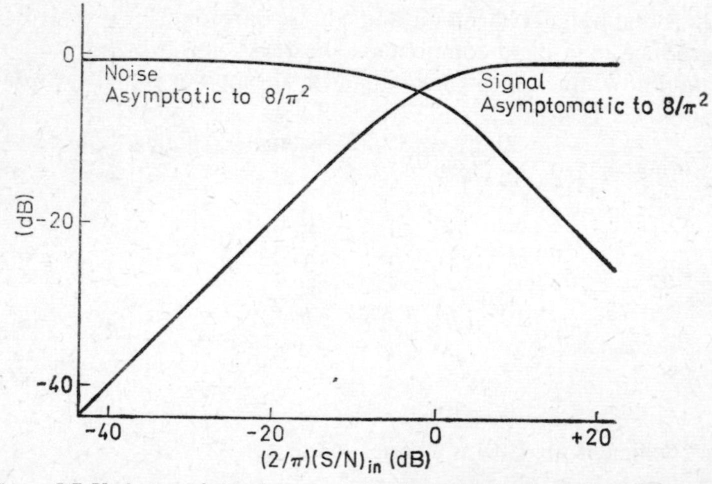

Figure 5.7 Variation of signal output and noise output with signal/noise ratio at the input for a hard limiter. Reproduced from Davenport[4] by courtesy of the I.E.E.E.

It is also possible to compute the value of the output noise power, N_{out}, as the input signal/noise ratio tends to zero, and when the input signal/noise ratio tends to infinity. When $(S/N)_{in} \to \infty$ equation 5.55 may be used, and it can be shown that only the first term in each of equations 5.58 and 5.59 is significant, and contribute equally to the final value of N_{out}. As $(S/N)_{in} \to 0$ all the contributions of the $(s \times n)$ type noise tend to zero relative to that of the $(n \times n)$ type, and the limit of equations 5.58 with 5.53 and 5.54 as $k \to \infty$, can be numerically computed. These results are

$$\left.\begin{array}{lll} \text{As} & (S/N)_{in} \to 0 & N_{out} \to 8/\pi^2 \\ \text{As} & (S/N)_{in} \to \infty & N_{out} \to 4/\pi^2 . (1)/(S/N)_{in} \end{array}\right\} \quad (5.61)$$

Combining equations 5.57 and 5.61

$$\left.\begin{array}{ll} (S/N)_{out} \approx 2(S/N)_{in} & \text{as} \quad (S/N)_{in} \to \infty \\ (S/N)_{out} \approx (\pi/4)/(S/N)_{in} & \text{as} \quad (S/N)_{in} \to 0 \end{array}\right\} \quad (5.62)$$

This example of the application of the characteristic function technique has yielded a number of very interesting principles. These can be summarised as follows:

(i) One signal in noise (or in a large number of other signals) cannot be suppressed by more than about one decibel $(\pi/4)$ even if the noise is very large. Here suppression is a measure of the change in signal/noise ratio from input to output.

(ii) The proportion of total output noise power that falls back across the fundamental band is $8/\pi^2$ for a hard limiter, if the input is noise alone. This happens to be the proportion of total output power that falls back across the fundamental band if the input consists of a single sinusoidal input.

(iii) The proportion of total output noise power across the kth harmonic (k odd) for a hard limiter is also identical to that for a sinusoid, viz., $8/(k\pi)^2$.

(iv) It might be concluded from (iii) and (ii) that as the number of sinusoids in the input to the hard limiter is increased until it is 'noiselike', the sum of the fundamental output sinusoids would remain a constant proportion, $(8/\pi^2)$, of the total output. However, the output noise is represented by equation 5.58 and only the first term is coherent with the input, the subsequent terms in $R_n^k(0)$, $(k>1)$, giving rise to $W_n(f) \overset{k}{*} W_n(f)$ spectra, which are of wider bandwidth than the input band, and therefore incoherent with it. The proportion of the output coherent with the input is therefore $(h_{01}^2)/1! = (2/\pi)$.

(v) If a large number of carriers is being transmitted through a nonlinear device, then, in general, a change in input level of one carrier will be transferred in some measure to all the other carriers. This follows from considering multiple carriers as noise in equation 5.57 for the case of low $(S/N)_{in}$. Thus if one carrier changes level (i.e. N_{in} changes) then S_{out} for each of the other carriers must also change.

(vi) The hard limiter is not the most distorting device. In fact it retains a lot of output $(2/\pi)$ coherent with the input. Above saturation a cubic low device will have a lower carrier to total noise ratio for Gaussian noise with one sinusoid than a hard limiter with the same input. This is given as an example for the reader to attempt at the end of this chapter.

(vii) The hard limiter can be used to enhance the signal/noise ratio at the output of a system, as transmission through a hard limiter of a signal which is large compared to the noise effectively removes the amplitude modulation component of the noise (which comprises half the noise power) so that the signal/noise ratio at the output is 3 dB (twice) better than that at the input.

5.3.5. Suppression of One Carrier by Another in a Hard Limiter

Equation 5.48 gives the amplitude of the noise and one sinusoidal signal and their harmonics at the output of a nonlinear device. This applies this to a number of sinusoids with random noise. Thus, this section effectively goes back to the start of equation 5.33 where the signal, or more accurately, its characteristic function, had still to be specified. This turns out to be a trivial extension of equation 5.33 as the characteristic function of a sum of sinusoids is the multiple of the individual characteristic functions (Section 1.3.7, equation 1.29). Thus

$$M_s(p_1 p_2 \tau) = M_{s1}(p_1 p_2) M_{s2}(p_1 p_2) \ldots M_{sn}(p_1 p_2) \quad (5.63)$$

$$= \sum_{m1=0}^{\infty} \sum_{m2=0}^{\infty} \ldots \sum_{mn=0}^{\infty} (\mathrm{e}_{m1}\mathrm{e}_{m2} \ldots \mathrm{e}_{mn}) \big(I_{m1}(p_1 A_1)\, I_{m2}(p_1 A_2)$$

$$\ldots I_{mn}(p_1 A_m)\big) \big(I_{m1}(p_2 A_1) . I_{m2}(p_2 A_2) \ldots I_{mn}(p_2 A_n)\big) \big(\cos (m_1 \omega_1 \tau)$$

$$\cos (m_2 \omega_2 \tau) \ldots \cos (m_n \omega_n \tau)\big)$$

Following the procedure outlined in Section 5.3.3, the amplitude of the various products is given by

$$h_{m1m2\ldots m_nk}$$

$$= \frac{(-1)^{\frac{k+m_1\ldots m_n-1}{2}}}{\pi} \int_0^\infty V^{k-1}\left\{\sum_{k=1}^{N} I_{mk}(VA_k)\right\} e^{\frac{-R_n(0)V^2}{2}}\, dV \quad (5.64)$$

To yield a final spectrum the 'cos' term in equation 5.63 will have to be transformed and will give delta function 'spikes' at frequencies $m_1\omega_1 \pm m_2\omega_2 \pm m_3\omega_3 \ldots m_n\omega_n$. By letting m_3 to m_n equal zero in equation 5.64 the amplitude of the products involving only ω_1 and ω_2 can be found. The carrier power is found by letting m_2 to m_n equal zero. Thus the carrier at ω_r is given by equation 5.64, with all $m_n = 0$ except m_r which $= 1$. Work in evaluating parts or all of equation 5.64 for various inputs has been performed by Davenport[4], Jones[8] and Shaft[9]. Figures 5.8, 5.9 and 5.10 show suppression of one carrier by another for combinations of other for two, three, four and an infinite number of carriers, for a hard limiter. From these figures the suppression of one carrier by another can be easily seen. Suppression is defined here as

$$\text{Suppression} = \frac{(S_A/S_B)_{\text{in}}}{(S_A/S_B)_{\text{out}}} \quad (5.65)$$

when signal $A(S_A)$ is said to have been suppressed relative to signal $B(S_B)$ by the value of the suppression. Figures (5.8)–(5.10) are intended to give an indication of how multiple carriers behave in a nonlinear device. The reader requiring more accurate knowledge should program equation 5.64 numerically. This data holds for only one (impractical) limiting characteristic. Before extending the range of characteristics, as done in the next section, some useful 'rules' regarding suppression will be listed here for general guidance.

1. 6 dB is the maximum suppression of one signal by another.
2. 1 dB is the maximum suppression of one signal by noise.
3. 3 dB is the maximum suppression of noise by a signal.
4. In a hard limiter the output at the fundamental frequencies of, and coherent with, the input for large numbers of carriers is 1 dB less than the fundamental power at the output when the input is one sinusoid.
5. Maximum suppression occurs with two carriers, when the strongest intermodulation product power can equal that of the weaker carrier.

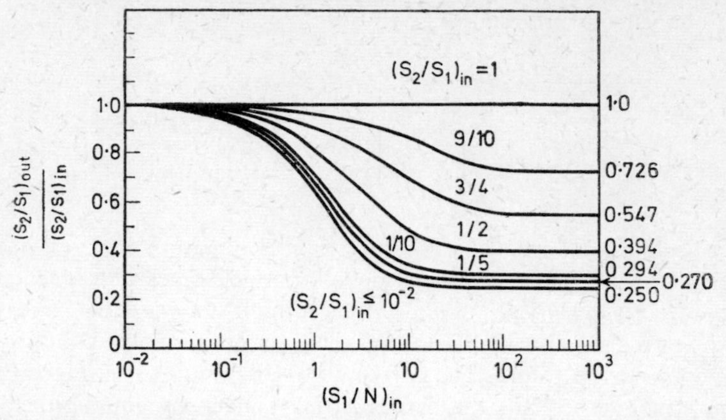

Figure 5.8. Two signal suppression. Reproduced from Jones[8] by courtesy of the I.E.E.E.

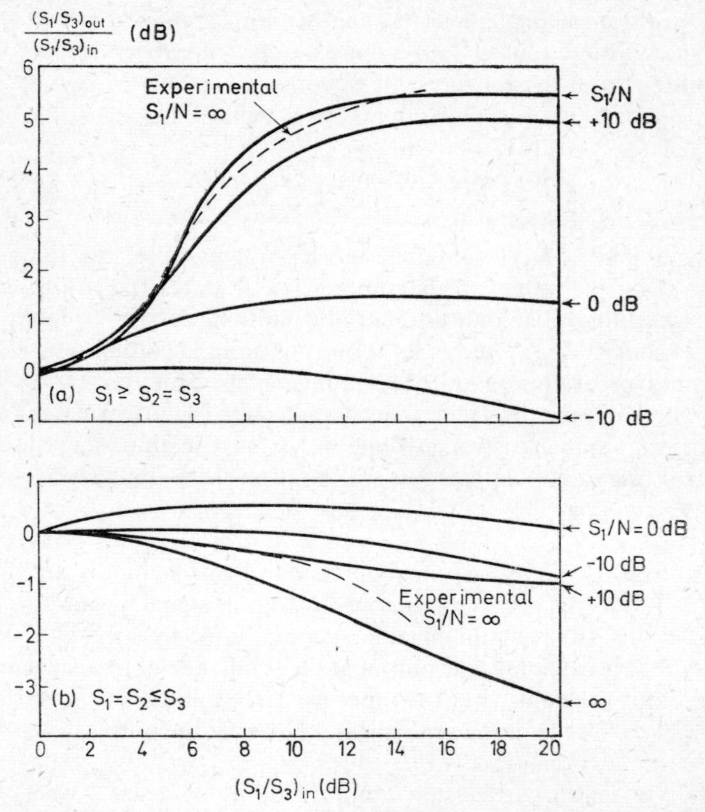

Figure 5.9 Three signal suppression. Reproduced from Shaft[9] by courtesy of the I.E.E.E.

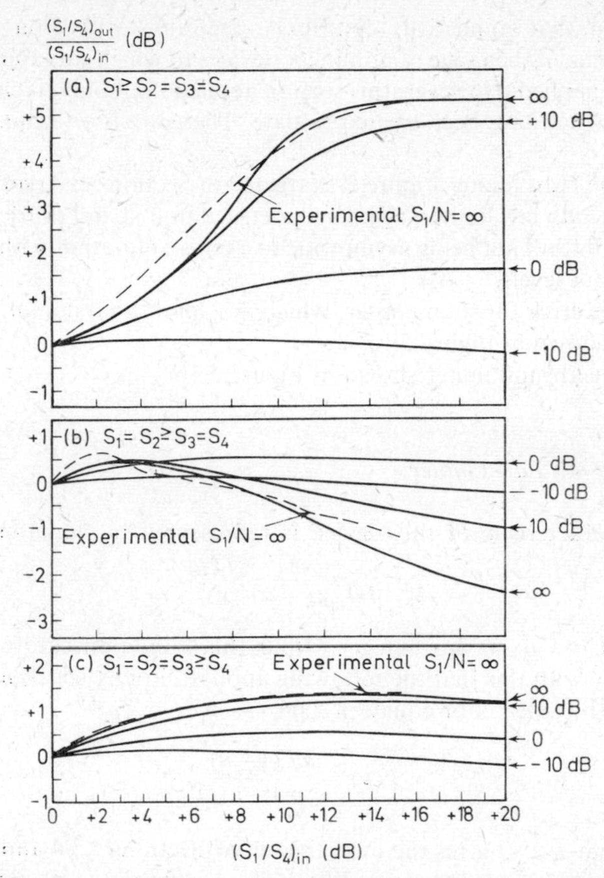

Figure 5.10 Four signal suppression. Reproduced from Shaft[9] by courtesy of the I.E.E.E.

6. If the power of two equal large carriers is significantly larger than the sum of the power of the remainder of the carriers and/or noise, then the power of the remaining signal and/or noise may be augmented with respect to the two large carriers at the input.

5.3.6. Nonlinear Transfer Characteristics

Basically speaking, equation 5.47 or its multi-carrier version 5.64 are the objective of the analysis, as these give output powers. Equation 5.47 can be applied to any characteristic whose Laplace transform is obtainable. In the past this had to be coupled with an ability to solve the resultant integral (cf. equation 5.52). It is to be expected that numerical (contour) integration will soon allow a widening of the range of nonlinear devices to which this approach may be applied. However, this section applied itself only to characteristics that have been treated to date. These are four in number,

1. The hard limiter Figure 5.1a treated in Section 5.3.4 and 5.3.5.
2. The nth law limiter, again symmetrical in first and third quadrants, but not being asymptotic to a constant output with large input levels.
3. The error function limiter, which is a smooth version of 1 and is shown in Figure 5.1c.
4. The abrupt limiter shown in Figure 5.1b.

(a) The nth *Law Limiter*

The characteristic of this device is represented by equation 5.32 where

$$f(V_{\text{in}}) = \alpha V_{\text{in}}^{n} \qquad (5.66)$$

where $n = 1/m$, m any integer. Note this condition carefully as it is only with this that the following applies. Strictly speaking this is an 'nth rooter'. For equation 5.66

$$F(p) = \frac{\alpha \Gamma(1+n)}{p(1+n)} \qquad (5.67)$$

Note that $n = 0$ gives the hard limiter of Sections 5.3.4 and 5.3.5 while $n = 1$ gives a linear amplifier. The coefficient h_{mk} is given by

$$h_{mk} = \frac{2\alpha\Gamma(1+n)\,(S/N)_{\text{in}}^{m/2}\;{}_1F_1\left[\dfrac{m+k-n}{2};\; m+1;\; -(S/N)_{\text{in}}\right]}{\left(\dfrac{R_n(0)}{2}\right)^{\frac{k-n}{2}}\Gamma(m+1)\,\Gamma\left(1-\dfrac{m+k-n}{2}\right)}$$

$m+k$ odd

(5.68)

Putting $n = 0$ gives equation 5.52a as expected.
Remember also (Appendix 5) that

$${}_1F_1(a;\; c;\; -z) = 1 - \frac{az}{c1!} + \frac{a(a+1)z^2}{c(c+1)2!}\cdots$$

$$m\Gamma(m) = \Gamma(m+1)$$

$$\Gamma(1/2) = \pi$$

equation 5.68 can now be used with equations 5.49, 5.58 or 5.59 to derive data corresponding to Figure 5.6 or equations 5.57, 5.61 or 5.62.

(b) The Error Function Limiter

The characteristic of this device is represented by equation 5.32 where

$$f(V_{in}) = \text{erf}\left(\frac{V_{in}}{a}\right) = \frac{1}{\sqrt{2\pi}} \int_{-\frac{\sqrt{(2)}V_{in}}{a}}^{\frac{\sqrt{(2)}V_{in}}{a}} e^{-z^2/2}\, dz \qquad (5.69)$$

The corresponding Laplace transform is

$$F(p) = \frac{1}{p}\, e^{\frac{a^2p^2}{2}} \qquad (5.70)$$

If equation 5.70 is applied to 5.47

$$h_{mk} = \frac{1}{2\pi j}\int p^{k-1}.I_m(pA)\, e^{\frac{(R_n(0)+a^2)p^2}{2}}\, dp \qquad (5.71)$$

Thus the noise term is augmented by a^2, which is a measure of the rate of increase of the limiter curve, see Figure 5.1c. For a hard limiter, i.e. $a = 0$, no noise augmentation takes place. h_{mk} is given by equation 5.68 where

$$(S/N)_{in} = \frac{S_{in}}{(R_n(0)+a^2)} \qquad (5.72)$$

The peculiar relationship between effective noise increase and the transfer characteristic is due to the close relationship of each to the Gaussian characteristic. Notice the apparent anomaly that the 'more nonlinear' device produces less noise—a fact referred to in (vi) of Section 5.3.4.

From (a) and (b) it is also possible to form more exotic transfer characteristics by convolving equations 5.69 and 5.66. Thus, of course, the Laplace transform of the resultant characteristic is the multiple of equations 5.67 and 5.70.

(c) The Abrupt Limiter

This transfer characteristic is shown in Figure 5.1b letting β in Figure 5.1b be unity (the output can be scaled up to suit practical values of amplification), then

$$f(V_{in}) = \begin{cases} V_{in} & \text{for} \quad V_{in} < a \\ a & \text{for} \quad V_{in} > a \end{cases} \tag{5.73}$$

$f(V_{in})$ can be expressed as

$$f(V_{in}) = \frac{2}{\pi} \int_0^\infty \sin V_{in}u . \sin au \frac{du}{u^2} \tag{5.74}$$

Only an input of Gaussian noise has been treated for this characteristic and an outline of the working will be given here, as it is pertinent to clipping of speech signals, and is used in Section 6.4.2. Basically, the problem is to find the ratio of the power of that part of the output coherent with the input to the rest of the output power.

Equation 5.74 corresponds to 5.27 where $F(u) = \dfrac{\sin au}{u^2}$. As equation 5.74 is formulated in terms of sin $(V_{in}u)$ and not $e^{jV_{in}u}$ the equation corresponding to equation 5.28 has four characteristic functions in it. Writing this out fully,

$$R_{out}(\tau) = \frac{4}{\pi^2} \int_0^\infty \int_0^\infty E\,(\sin V_{in}u . \sin V_{in}(\tau)v) \left(\frac{\sin au \sin av}{u^2 v^2} \right) du\, dv \tag{5.75}$$

where $E[\quad]$ as usual indicates expectation or average. Using equation 5.21

$$E[\sin V_{in}u . \sin V_{in}(\tau)v] = e^{\frac{-(R_{in}(0).u^2 + R_{in}(0)v^2)}{2}} \sinh\,(R_{in}(\tau)uv) \tag{5.76}$$

From equations 5.76 and 5.75

$$R_{out}(\tau) = \frac{4}{\pi^2} \int_0^\infty \int_0^\infty e^{-R_{in}(0)\frac{u^2+v^2}{2}} \times \sinh R_{in}(\tau)\, uv \frac{\sin au \sin av}{u^2 v^2} du . dv \tag{5.77}$$

Expanding equation 5.77 and interchanging the order of summation and integration,

$$R_{out}(\tau) = \frac{4}{\pi^2} \sum_{m=0}^{\infty} \frac{R_m(\tau)^{2m+1}}{(2m+1)!} \int_0^{\infty} u^{2m-1}\, e^{\frac{-R_{in}(0)u^2}{2}} \sin au\, du$$

$$\times \int_0^{\infty} v^{2m-1}\, e^{\frac{-R_n(0)v^2}{2}} \sin av\, dv \qquad (5.78)$$

The wanted signal is given by $m = 0$ in (5.78).
Thus,

$$R_{out}(\tau)_{wanted} = \frac{4}{\pi^2} R_{in}(\tau) \left[\int_0^{\infty} \frac{1}{u}\, e^{\frac{-R_n(0)u^2}{2}} \sin au\, du \right]^2$$

$$= R_{in}(\tau)\, \mathrm{erf}^2 \left(\frac{a}{\sqrt{(2R_n(0))}} \right) \qquad (5.79)$$

The total output power $R_{out}(0)$ is given by

$$R_{out}(0) = \int_{-\infty}^{\infty} x^2 p(x)\, dx \qquad (5.80)$$

In this case $p(x)$, from the transfer characteristic, is

$$p(x) = \begin{cases} \frac{1}{2}\, \delta(x+a)\, [1 - \mathrm{erf}\; a/\sqrt{(2R_n(0))}] \\ f(x) \\ \frac{1}{2}\, \delta(x-a)\, [1 - \mathrm{erf}\; a/\sqrt{(2R_n(0))}] \end{cases} \qquad (5.81)$$

From equations 5.80 and 5.81, and normalising to $R_n(0) = 1$, using Reference 10, Equation 3.12

$$R_{out}(0) = a^2[1 - \mathrm{erf}\; (a/\sqrt{2})] + 2/\sqrt{(2\pi)} \int_0^a x^2\, e^{-x^2/2}\, dx$$

$$= a^2 - \sqrt{(2/\pi)}a\, e^{-a^2/2} + (1-a^2)\, \mathrm{erf}\; (a/\sqrt{2}) \qquad (5.82)$$

so that finally the signal/noise ratio at the output of an abrupt limiter with an input consisting of a noiselike signal of unity power is

$$\frac{\mathrm{erf}^2\; (a/\sqrt{2})}{a^2 - \sqrt{(2/\pi)}a\, e^{-a^2/2} + (1-a^2)\, \mathrm{erf}\; (a/\sqrt{2}) - \mathrm{erf}^2\; (a/\sqrt{2})} \qquad (5.83)$$

and this result is used in Section 6.4.2 as equation 6.1.

5.4. CLASS III. DEVICES REPRESENTED BY ASYMMETRICAL CHARACTERISTICS

The devices considered in this section have transfer characteristics as shown in Figure 5.2. The three devices considered here are of the half wave rectifier type, i.e. for negative inputs the output is zero. These are more general devices than those considered in Section 5.3. in that the odd function characteristics (which are symmetrical) merely double the odd harmonic output and eliminate the even harmonic output. Similarly, even characteristics such as that for $f_- = f_+$ in Figure 5.2a (full wave rectifier) are not treated explicitly in this section as the even symmetry of the device doubles the even harmonic and annihilates the odd, compared to the corresponding half wave rectifier.

Only an input of a modulated sinusoidal signal with random noise is considered in this section.

The characteristic function method of Section 5.3 is used so that the objective is to solve equation 5.38 for a particular characteristic and input signal. The reader interested in detail should consult sections 5.3.1–5.3.3. Equation 5.38 becomes equation 5.46 for an input of one sinusoid and noise, and the amplitude of each component of the input is given by equation 5.47. The corresponding amplitude when the input consists of a number of input sinusoids is equation 5.64. Only one sinusoid will be treated here, as in this case results appear in closed form, but the extension to a number of sinusoids is not complicated, but is tedious in that a computer will be needed. The three characteristics will be treated in turn.

5.4.1. The Biased *n*th law Rectifier[11]

$$
\begin{aligned}
f+(V_{\text{in}}) &= \alpha(V_{\text{in}}-b_0)^n \qquad V > b_0, \quad n > 0 \\
&= 0 \qquad\qquad\qquad\quad\; V < b_0
\end{aligned}
\tag{5.84}
$$

Therefore

$$
F(p) = \frac{\alpha\Gamma(n+1)\,\mathrm{e}^{-pb_0}}{p^{(n+1)}}
\tag{5.85}
$$

The integral of equation 5.46 must now be found. It is not intended to give details of the calculations here. These will be found in an excellent paper by Middleton[11], for each of the three devices discussed here.

Let the amplitude of the sinusoid be $A(t)$, (where in Section 5.3.4 this was taken to be a constant A), then define two new pa-

rameters,
$$\left(\frac{S}{N}\right)_{\text{in}} = \frac{[A(t)]^2}{2R_n(0)}$$
$$b = \frac{b_0}{(R_n(0))^{1/2}} \tag{5.86}$$

If $b < 1$ use the following to compute the h_{mk}

$$h_{mk}(t) = \frac{\alpha\Gamma(n+1)\,\mathrm{e}^{-j(m+k)(\pi/2)}}{2m!}\left(\frac{R_n(0)}{2}\right)^{(n-k)/2}\cdot\left(\frac{S}{N}\right)_{\text{in}}^{m/2}$$
$$\sum_{l=0}^{\infty}\frac{(-1)^l 2^{l/2} b^l {}_1F_1[(l+m+k-n)/2;\ m+1;\ -(S/N)_{\text{in}}]}{l!\,\Gamma[(2+n-l-m-k)/2]} \tag{5.87}$$

If b is larger than unity, use the following.

$$h_{mk}(t) = \frac{\alpha\Gamma(n+1)}{2}\mathrm{e}^{-j(m+k)\pi/2}\left(\frac{R_n(0)}{2}\right)^{(n-k)/2}\cdot\left(\frac{S}{N}\right)_{\text{in}}^{m/2}$$
$$\times\sum_{l=0}^{\infty}\frac{(S/N)_{\text{in}}}{l!(l+m)!}\left\{\frac{{}_1F_1[(2l+m+k-n)/2;\ \tfrac{1}{2};\ -b^2/2]}{\Gamma[(2+n-2l-m-k)/2]}\right.$$
$$\left.-\sqrt{2}\,b\,\frac{{}_1F_1[(2l+m+k-n-1)/2;\ 3/2;\ -b^2/2]}{\Gamma[(1+n-2l-m-k)/2]}\right\} \tag{5.88}$$

The difference between these equations lies in their rates of convergence, and they can be used to obtain answers in a number of limiting cases of interest. For example the noise component can tend to zero, giving results pertinent to a Schottky barrier diode the noise can be allowed to predominate and the signal tend to zero, or the bias can tend to zero so giving half wave rectification; signal and noise can both exist with either predominant.

For a half wave rectifier, ($b_0 = 0$), therefore

$$h_{mk}(t) = \frac{\alpha\Gamma(n+1)}{2m!}\,\mathrm{e}^{-j(m+k)(\pi/2)}\left(\frac{R_n(0)}{2}\right)^{n-k/2}\left(\frac{S}{N}\right)_{\text{in}}^{m/2}$$
$$= \frac{{}_1F_1\left[\left(\frac{k+m-n}{2}\right);\ m+1;\ -\left(\frac{S}{N}\right)_{\text{in}}\right]}{\Gamma[(2+n-m-k)/2]} \tag{5.89}$$

For noise alone, $b_0 > 0$

$$h_{0k}(t) = \frac{\alpha\Gamma(n+1)\,\mathrm{e}^{-j(k\pi/2)}}{2}\left(\frac{R_n(0)}{2}\right)^{(n-k)/2}$$
$$\times\frac{{}_1F_1[(k-n)/2;\ \tfrac{1}{2};\ -b^2/2]}{\Gamma([2+n-k)/2]}$$
$$-b\sqrt{2}\,\frac{{}_1F_1[(1+k-n)/2;\ \tfrac{3}{2};\ -b^2/2]}{\Gamma([1+n-k)/2]} \tag{5.90}$$

When noise component is zero and bias is zero, then the amplitudes of the harmonics are given by

$$h_{m0} = \frac{\dfrac{\alpha\Gamma(n+1)}{2^{n+1}}\, e^{-j(m\pi)/2} A_0(t)^n}{\Gamma((2+n+m)/2)\,.\,\Gamma((2+n-m)/2)} \tag{5.91}$$

Other asymptotes of equations 5.87 and 5.88 are given in Reference 11.

5.4.2. The Half Wave Rectification of Random Noise[11]

This is really a special case of Section 5.4.1 where the characteristic is

$$\begin{aligned} F_+(V_{in}) &= \alpha V^n \quad V>0 \quad n>0 \\ f_-(V_{in}) &= 0 \end{aligned} \tag{5.92}$$

In this case the discussion touches on wideband and narrowband noise, i.e. includes spectrum information. To find the spectrum, as usual, first the autocorrelation function must be found. Using equations 1.15 and 1.33 (as the noise is assumed to be Gaussian) then with 5.92 and remembering that

$$\begin{bmatrix} \mu_{11} & \mu_{12} \\ \mu_{21} & \mu_{22} \end{bmatrix} = \begin{bmatrix} R_n(0) & R_n(\tau) \\ R_n(\tau) & R_n(0) \end{bmatrix} \tag{5.93}$$

then

$$R(\tau) = \frac{\alpha^2 R_n(0)^{n+1}}{2\pi(R_n(0)^2 - R_n^2(\tau))^{1/2}} \int_{-\infty}^{\infty}\int_{-\infty}^{\infty} (x_1 x_2)^n \times \exp\left[\frac{-R_n(0)(x_1^2+x_2^2)+2R_n(\tau)x_1x_2}{2(R_n(0)^2-R_n(\tau)^2)}\right] dx_1\, dx_2 \tag{5.94}$$

Equation 5.94 can be integrated using polar co-ordinates, and after some manipulation, yields the following form useful for transforming to find the spectrum:

$$R(\tau) = \frac{\alpha^2 2^n}{4\pi} R_n(0)^n \sum_{k=0}^{\infty} \left[\frac{(-n/2)_k^2\, \Gamma[(n+2)/2]^2\, 2^{2k}\,.\,\varrho(\tau)^{2k}}{(2k)!}\right] + \sum_{k=0}^{\infty}\left[\frac{((1-n)/2)_k^2\, \Gamma[n/2+1]\,.\,2^{2k+1}\,.\,\varrho(\tau)^{2k+1}}{(2k+1)!}\right] \tag{5.95}$$

From equation 5.95 putting $n = 1$ gives the half wave linear rectifier result,

$$R(\tau)_{n=1} = \frac{\alpha^2 R_n(0)}{2\pi}\left\{\varrho\left(\sin^{-1}\varrho+\frac{\pi}{2}\right)+(1-\varrho^2)^{1/2}\right\}$$

$$\text{where}\quad \varrho = \frac{R_n(\tau)}{R_n(0)}$$

and for $n = 2$ the half wave quadratic detector,

$$R(\tau)_{n=2} = \frac{\alpha^2 R_n(0)^2}{2\pi}\left\{\left(\frac{\pi}{2}+\sin^{-1}\varrho\right)(1+2\varrho^2)+3r(1-\varrho^2)^{1/2}\right\} \tag{5.97}$$

The effect of narrowband noise entering the rectifier may be studied by using a narrowband spectrum version of $R_n(\tau)$ (Equation 5.95). Some simplification of equation 5.95 may be obtained by writing $R_n'(\tau)$ for $R_n(\tau)$ where

$$R(\tau) = R_n(0).\varrho(\tau).\cos\omega_c(\tau) \tag{5.98}$$

and expanding the $R_n'(\tau)$ term in equation 5.95 to allow consideration of those components which fall back into the spectral region under consideration, in the manner of Sections 5.2.1 and 5.2.3. In fact, this approach could be considered as a technique leading up to the correct transfer characteristic coefficients for a half wave device for use in such equations as 5.13 and 5.14.

5.4.3. Small Signal Detection[11]

Here the input amplitudes are small so that cut-off and limiting of the signal can be ignored. The more general case of both odd and even powers is considered here, as opposed to the odd powers only as in Section 5.2.3. Thus

$$V_{out} = \sum_{k=0}^{\infty} \alpha_k V^k \tag{5.99}$$

For noise alone as the input the correlation function of the lth harmonic is given by *l even*

$$R_e(\tau) = e_{l/2}\frac{\varrho^l\cos l\omega_c t}{l!}\sum_{j=0}^{\infty}\sum_{k=0}^{\infty}\alpha_j\alpha_k(2R(0))^{\frac{j+k}{2}}\left(\frac{1-j}{2}\right)_{l/2}$$

$$\left(\frac{1-k}{2}\right)_{l/2}\times\Gamma\left(\frac{j+1}{2}\right)\Gamma\left(\frac{k+1}{2}\right)\,{}_2F_1\left(\frac{l-j}{2},\frac{l-k}{2};l+1;\,\varrho^2\right)$$

j, k even

(5.100)

and for *l odd*

$$R_e(\tau) = \frac{2\varrho^l \cos l\omega_c t}{l!} \sum_{j=0}^{\infty} \sum_{k=0}^{\infty} \alpha_j \alpha_k (2R_n(0))^{\frac{j+k}{2}} \left(\frac{1-j}{2}\right)_{\frac{l-1}{2}}$$

$$\left(\frac{1-k}{2}\right)_{\frac{l-1}{2}} \Gamma\left(\frac{j}{2}+1\right) \Gamma\left(\frac{k}{2}+1\right) {}_2F_1\left(\frac{l-j}{2}, \frac{l-k}{2}; l+1; \varrho^2\right)$$

j, *k*. odd

(5.101)

where $${}_2F_1(\alpha, \beta; \gamma; x) = 1 + \frac{\alpha\beta}{\gamma} \frac{x}{1!} + \frac{\alpha(\alpha+1)\,\beta(\beta+1)}{\gamma(\gamma+1)} \frac{x^2}{2!} + \cdots$$

$$\text{and } (\alpha)_n = \alpha(\alpha+1)\ldots.(\alpha+n-1)$$

It is also possible to obtain answers for a sinusoid and noise for simple cases (quadratic polynomial equation 5.99) and these are given in very general form in Reference 11, section 8(*b*) and Reference 6, equation 4.10.1. and 4.10.3.

Problems

5.1. Prove that the expectation of cos $m\Theta$ times cos $n\Theta$ is zero unless $m = n$.

5.2. Show that for very low signal/noise ratios at the input of a hard limiter, of maximum output 1 V, the signal output is $2/\pi$ times the input signal/noise ratio. As the signal becomes very large relative to the noise show that the output signal power tends to $8/\pi^2$. Show that the noise output in this case tends to $4\pi.\left(\frac{N_{in}}{S_{in}}\right)$.

5.3. A narrow band f.d.m.-a.m. telephony signal of Gaussian noise characteristics input is applied to a hard limiter, limiting at ± 1 V. What is the output signal/noise power falling onto the fundamental band if the r.m.s. voltage is (a) 0·1 V (b) 0·361 V (c) 1 V?

Ans. (a) 3·66 (b) 3·66 (c) 3·66

5.4. If the signal of problem 5.3 is applied to a nonlinear characteristic given by

$$V_{out} = 36V_{in} - 7V_{in}^3$$

what is the output signal/noise ratio in the fundamental band if the r.m.s. voltage of the signal is (a) 0·1 (b) 0·361 V (c) 1 V? Assume all values normalised to 1 Ω.

Ans. (a) 47·6 dB (b) 27·2 dB (c) 0·1 dB

5.5. Prove the relationships for the fourth order moments given in Section 5.2.3 for a power law device (see equation 5.10 *et seq.*).

References

1. WASS C. A. A., 'A Table of Intermodulation Products, *J.I.E.E.*, p 95, Pt III, p. 31 (1948)
2. LANNING J. H. and BATTIN, R. H., *Random Processes in Automatic Control*, McGraw Hill Book Co. Inc. (1956)
3. BENNETT W. R., 'Distribution of the Sum of Randomly Phased Components', *Quarterly Appl. Math.*, **5**, 385, (1948)
4. DAVENPORT W. B. JR., Signal to Noise Ratios in Bandpass Limiters. *J. Appl. Phys.*, **24**, No. 6, 720 (June 1953)
5. DWIGHT H. B., *Mathematical Tables of Elementary and some Higher Mathematical Functions*, Dover Publications Ltd., New York.
6. RICE S. O., 'Mathematical Analysis of Random Noise, *B.S.T.J.*, **23**, 282 (1944), and **24**, 46 (1945)
7. CAHN C. R., 'Crosstalk due to Finite Limiting of Frequency Multiplexed Signals'. *Proc. I.R.E.*, 53 (Jan 1960)
8. JONES J. J., 'Hard Limiting of Two Signals in Random Noise, *I.E.E.E. Trans. Inf. Theory*, 34 (Jan 1963)
9. SHAFT P. D., 'Limiting of Several Signals and its effect on Communication System Performance, *I.E.E.E. Trans. Commun. Theory*, **COM-13** No. 4, 504 (Dec. 1965)
10. DAVENPORT W. B. and ROOT W. L. *Random Signals and Noise*, Chapters 12 and 13, McGraw Hill Book Co. Inc. (1956)
11. MIDDLETON D., 'Some General Results in the Theory of Noise through Nonlinear Devices'. *Quarterty Appl. Math.*, **5** (Jan 1948) p 445.
12. WESTCOTT R. J., 'Investigation of multiple f.m./f.d.m. carriers through a satellite T.W.T. operating near to saturation', *Proc. I.E.E.*, 726 (June 1967)

Chapter 6

NOISE IN PULSE MODULATION SYSTEMS

6.1. INTRODUCTION

In Chapter 1, it was indicated that there are two broad classifications of methods of signal modulation, the continuous carrier modulation methods (amplitude modulation and angle modulation) and pulse modulation methods. As the names imply, in the continuous methods the modulated signal is present continuously while in pulse methods the signal is only present at discrete times. This property of pulse modulation allows systems to be multiplexed in time (see Chapter 1), which is used to advantage in many systems such as telephone exchanges for example.

The process of continuous modulation introduces no noise into a communication channel since the continuously varying or analogue signal may continuously vary a parameter of the carrier (i.e. its amplitude or its phase), and hence there is no loss of information. This statement is not strictly true since in a practical system there must be some limit on parameters such as bandwidth in the case of frequency modulation or amplitude in the case of amplitude modulation and this will of course introduce some distortion. There are some modulation methods however which actually introduce noise into the modulation process. These pulse modulation methods will be referred to as digital modulation methods and the types which introduce no noise will be termed analogue pulse modulation methods.

Digital modulation methods have the advantage that the modulated signal takes on only a limited number of definite forms i.e. the signal can be only in a finite number of different states. These states can be known to the receiver and so instead of a direct measurement of some parameter which occurs in analogue modulation, there can be a comparison between the incoming signal (which may be distorted by additional noise from interference for example) with the known possibilities that can occur. Thus the concept of a receiver making a decision between one of many possibilities arises, and this aspect of digital modulation is discussed further in Sections 6.6 and 6.7.

In the following analysis the Gaussian noise model of the modulating signal will be used where necessary, which, as explained in Chapter 3, is not very representative of a single talker but rather a block of multiplex speech channels. This model is used simply because it lends itself to analysis and serves to show how noise calculations can be made for the various systems.

6.2. SAMPLING OF SIGNALS

Sampling of the amplitude of a signal at discrete points in time is the basis of all pulse modulation methods and the following theorem relates the sampling rate to the maximum frequency of the signal to be sampled i.e. its bandwidth.

If a waveform has frequencies in its spectrum extending from a lower frequency limit to an upper frequency limit f_m Hz it is possible to convey all the information in that waveform by $2f_m$ or more equally spaced samples per second of the amplitude of the waveform.

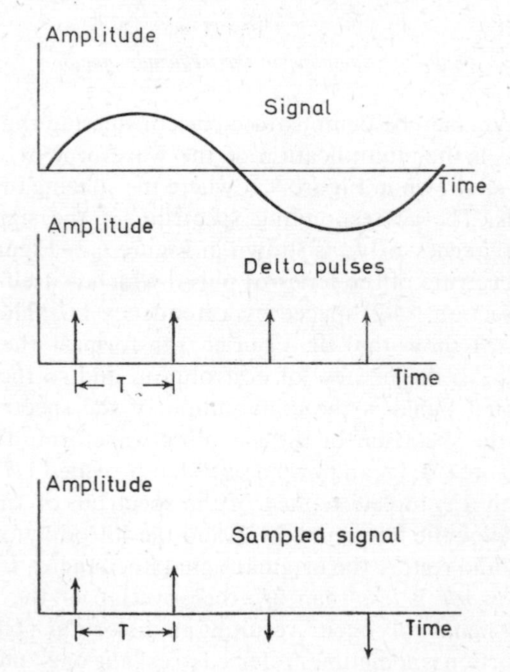

Figure 6.1 Sampling of a signal by delta pulses

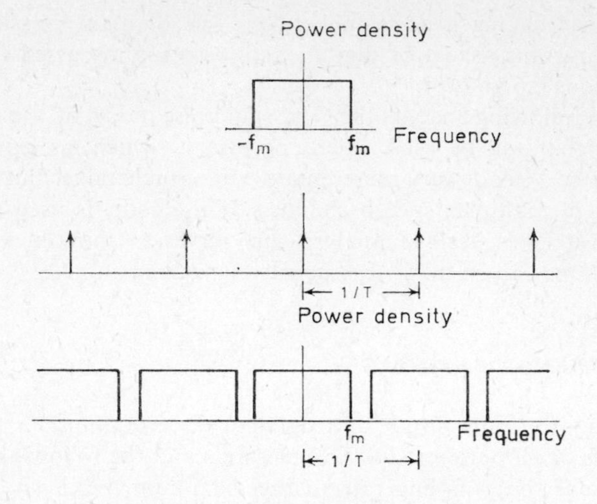

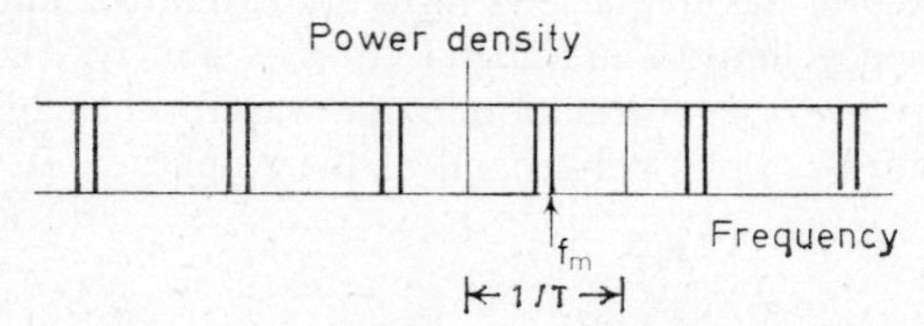

Figure 6.2 Sampling in the frequency domain

The theorem may be demonstrated by considering that the sampling process is the multiplication of the waveform by a series of delta pulses as shown in Figure 6.1, where the spacing of the pulses is T seconds. The corresponding spectrum of the signal with a maximum frequency of f_m is shown in Figure 6.2. Figure 6.2 also shows the spectrum of the series of pulses which is itself a series of pulses (see Section 1.4.2) spaced by a frequency $1/T$. The theorems of Appendix 1 show that the Fourier transform of the operation of multiplying is the process of convolution and so the spectrum of the sampled signal is the convolution of the spectrum of the signal with the spectrum of the sampling waveform. This is also shown in Figure 6.2. It can now be seen that provided $1/T$ is greater than $2f_m$ then a complete replica of the spectrum of the sampled signal lies below the frequency $1/2T$ and the introduction of a low pass filter would restore the original signal unchanged. If, however, the frequency $1/T$ is less than $2f_m$ then overlap of the spectra of the sampled signal will occur resulting in distortion. This mechanism of distortion is sometimes referred to as 'aliasing' and is shown in Figure 6.2.

If a signal with a background of noise is sampled, the signal/noise ratio compared with the original signal/noise ratio depends on the bandwidth of the noise. Obviously if the noise occupies the same bandwidth as the original signal then the noise will be sampled in exactly the same way as the signal and the signal/noise ratio will be unchanged by the sampling process. In digital modulation a case arises, however, where the background noise has a

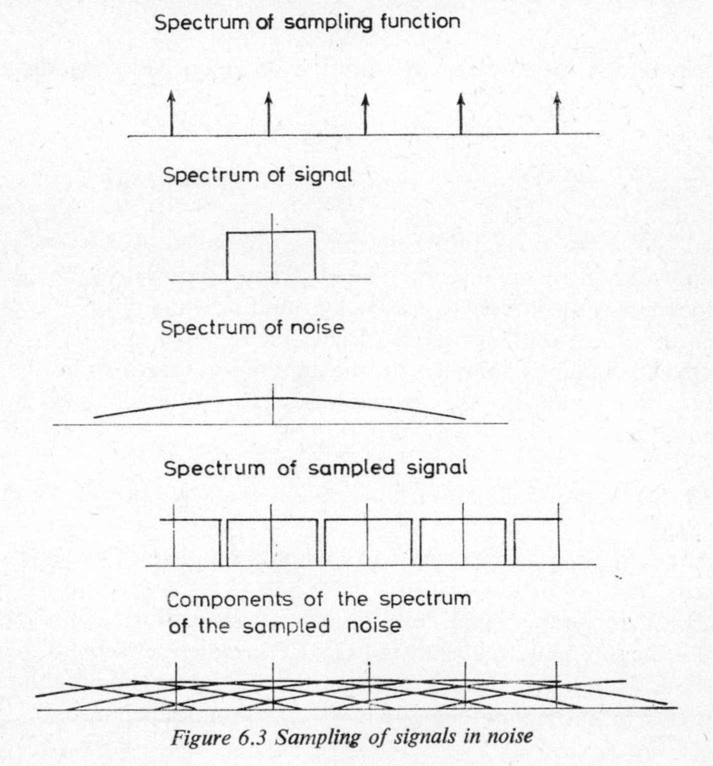

Figure 6.3 Sampling of signals in noise

much larger bandwidth than the signal. The effect of sampling now is shown in Figure 6.3 where it can be seen that the noise spectrum overlaps considerably and that if the sampling rate is only just greater than $2f_m$ then to a first approximation the total power of the noise all falls within the signal bandwidth. Thus the signal/noise ratio is degraded by the sampling process. More generally it can be seen that provided the noise spectrum at the input is considerably wider than the sampling rate, the total noise power at the input is evenly distributed over each output frequency band of width $1/T$. If, therefore, the total noise power at the input is N_{in} and the signal power is S_{in} the signal/noise ratio in a narrow band centred on the frequency f at the output is given approximately by

$$\left(\frac{S(f)}{N_0(f)}\right)_{out} = \frac{S_{in}}{2f_m N_{in} T} \quad 1/T \geqslant 2f_m$$

If $1/T = 2f_m$, then

$$\left(\frac{S(f)}{N_0(f)}\right)_{out} = \frac{S_{in}}{N_{in}}$$

even though the noise at the input is of much wider bandwidth than the signal.

Example 6.1

An f.d.m. block of twelve telephone channels, each of 4 kHz bandwidth, is to be pulse modulated. The total power of the f.d.m. block is P watts and there is a background of white noise of power density of $P/6$ watts per MHz (single sided).

If the signal is sampled at the minimum rate for no signal distortion, calculate the signal/noise ratio in a channel after demodulation if:

A. (a) A pre-modulation filter of low-pass bandwidth 48 kHz is used

(b) A pre-modulation filter of low-pass bandwidth 192 kHz is used.

B. Calculate the signal/noise ratio in a channel for cases (a) and (b) if the sampling frequency is 192×10^3 samples per second.

C. Calculate the signal/noise ratio in the top baseband channel for case (a) if the sampling rate is 90×10^3 samples per second.

Solution

The bandwidth of the single sided power spectrum of the f.d.m. block of channels is $12\times4 = 48$ kHz. Therefore the minimum sampling rate for no signal distortion is 2×48 kHz $= 96$ kHz. A. (a) In this case the background noise occupies the same bandwidth as the signal and hence the S/N at the output of a demodulator will be the same as that of the input i.e.

$$S/N = \frac{P}{(P/6)\times10^{-3}\times48} = 125 \qquad (21\text{ dB})$$

A. (b) In this case the noise spectrum has a bandwidth of four times the signal bandwidth and the two sided spectra are shown in Figure 6.4. The power density of the signal is $P/(2\times48)$ W/kHz and the power density of the noise is $P/(6\times2)$W/MHz. Referring to Figure 6.2, the components of the convolution are shown in

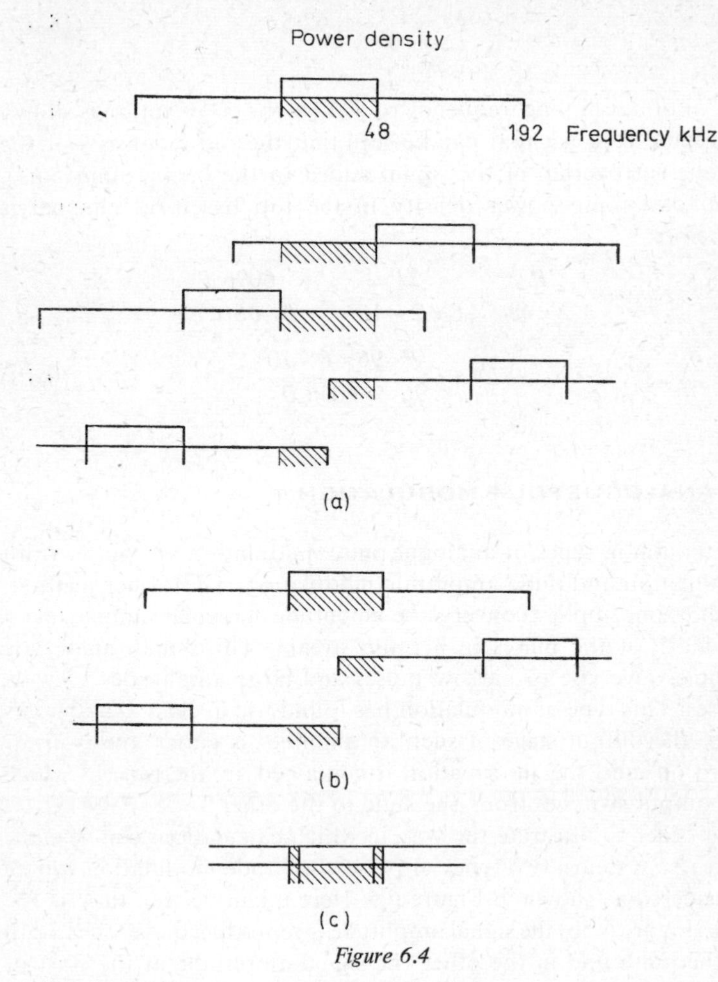

Figure 6.4

Figure 6.4a and it can be seen that the noise falling in the signal bandwidth is now four times as great, i.e.

$$S/N = \frac{125}{4} = 31{\cdot}25 \qquad (15\text{ dB})$$

B. (a) Changing the sampling frequency to 192 kHz will not

change the situation in case (a) since the signal and noise occupy the same bandwidth.

B. (b) For a sampling frequency of 192 kHz Figure 6.4b shows that the noise is reduced by half, i.e.

$$S/N = \frac{125}{2} = 62{\cdot}5 \qquad (18\text{ dB})$$

C. For a sampling frequency of 90 kHz the convolution is shown in Figure 6.4c where it can be seen that the noise consists of the incoherent overlap of the signal added to the background noise. The total noise power density in the top baseband channel is therefore,

$$\frac{P}{2\times 48} + \frac{2P}{6\times 2\times 10^3} = \frac{6096\,P}{96\times 6\times 10^3}$$

$$\therefore\ S/N = \frac{P}{96}\,\frac{96\times 6\times 10^3}{6096\,P} \approx 1 \qquad (0\text{ dB})$$

6.3. ANALOGUE PULSE MODULATION

The common types of analogue pulse modulation are pulse width modulation and pulse amplitude modulation. The former method, as the name implies, converts the amplitude of regular samples taken to width of the pulses in a pulse stream. Thus small amplitude samples give rise to narrow pulses and large amplitudes to wide pulses. This type of modulation has found use in class D amplifiers since the output stage of such an amplifier is either 'hard on' or 'hard off' and the information is contained in the time at which the output switches from one state to the other.

In order to illustrate the way in which calculations can be made for these systems, two types of pulse amplitude modulation will be considered as shown in Figure 6.5. Here it can be seen that in one case (Figure 6.5b) the signal amplitude is reproduced over the length of the pulse and in the other the signal amplitude at the start of the pulse is prolonged for the width of the pulse (Figure 6.5c).

The properties of the Fourier transform (Appendix 1) may be used to calculate the form of the spectrum for these two cases by first considering how the waveforms are made up in the time domain and then converting the operations to the frequency domain. Thus in Figure 6.6a it can be seen that the final time waveform is made by simply multiplying the analogue signal by a pulse waveform of unit height. Now since the operation of multiplication in

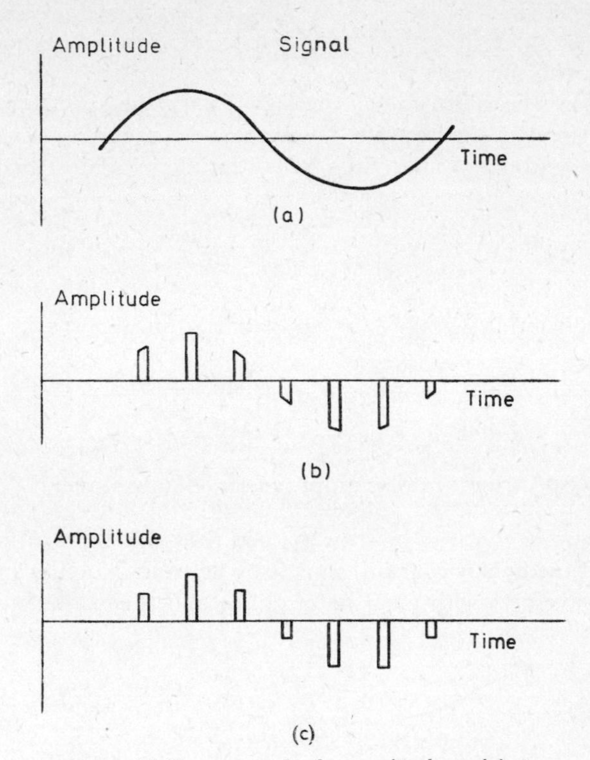

Figure 6.5 Two types of pulse amplitude modulation

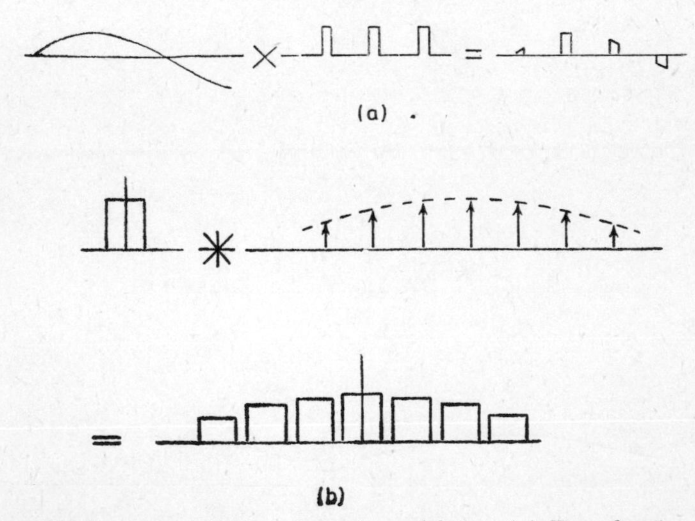

Figure 6.6 Formation of pulse amplitude modulation. (a) Time domain; (b) frequency domain

the time domain is transformed to convolution in the frequency domain, the spectrum for this type of modulation is obtained by convolving the spectrum of the signal with the spectrum of the pulse train as in Figure 6.6b. The amplitude spectrum for this type of pulse modulation may thus be written down almost by inspection,

$$A(f) \quad * \quad \frac{\sin \pi\tau f}{\pi\tau f}\,\frac{\tau}{T}\sum_n \delta(f-n/T)$$

↓ Spectrum of signal. ↓ Spectrum of pulse train.

$$= \frac{\tau}{T}\sum_n A(f-n/T)\,\frac{\sin n\pi\tau/T}{n\pi\tau/T}$$

↓

Spectrum of pulse amplitude modulated signal.

In Figure 6.7a it can be seen that for the other form of modulation that the final waveform is made by first sampling the analogue signal waveform with ideal delta pulses and then broadening the

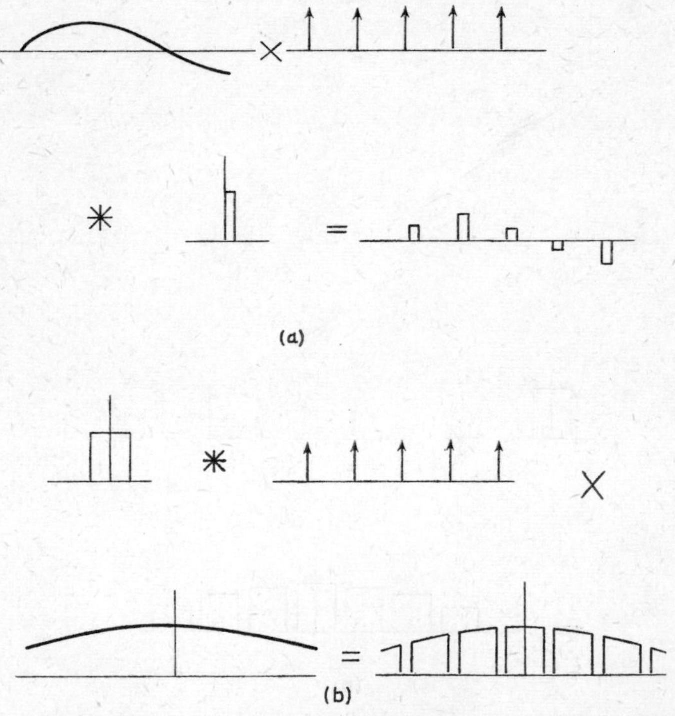

Figure 6.7 Formation of pulse amplitude modulation. (a) Time domain; (b) frequency domain

delta pulses out into pulses of a practical width whose height corresponds to that at the time of sampling i.e. a sample and hold arrangement. This broadening out is achieved mathematically by convolving the sampled waveform with a single rectangular pulse of unit height. In the frequency domain therefore the spectrum of the analogue waveform is first convolved with a train of delta pluses and the result is multiplied by the spectrum of the rectangular pulse. This is shown pictorially in Figure 6.7b.

6.4. PULSE CODE MODULATION [1]

6.4.1. Description of Modulation Process

Pulse code modulation (p.c.m.) is a digital modulation method which relies upon some loss of information of the original signal being incorporated. This loss of information manifests itself as quantising noise.

In pulse code modulation the analogue signal is first sampled and the amplitude of each sample is used to encode a stream of pulses over a fixed period of time. The coding used is often binary because it is often easier to send a signal which has two states (e.g. on and off) rather than more than two states. Thus during the fixed time interval corresponding to one sample a fixed number of pulses are sent and these make up the code word defining the amplitude of the sample. If there are three pulses corresponding to each sample (for example) then the number of different combinations that can be sent is 2^3 i.e. 8. The magnitudes corresponding to these combinations are shown in Figure 6.8. Thus only eight possibilities of amplitude are possible and the actual amplitude of any particular sample must be altered so that it is equal to the nearest permissible amplitude for transmission. There is therefore an error between the actual level of the signal and the signal transmitted, and the process of making the amplitudes of the samples

Pulse code	Sample amplitude
1 1 1	3
1 1 0	2
1 0 1	1
1 0 0	0
0 1 1	−1
0 1 0	−2
0 0 1	−3

Figure 6.8 Binary Coding

correspond to discrete levels is termed quantising, and the noise that the errors introduce is called quantising noise. Obviously the magnitude of the errors depends upon the number of pulses which comprise a word and this in turn affects the bandwidth of the system.

In analogue carrier modulation systems the modulation parameters (such as depth of modulation or frequency deviation) are generally determined by the nature of the noise in the communication channel over which the information is to be sent (e.g. cross-talk in cables and interference in radio links). In digital modulation systems these parameters are also chosen with regard to the noise caused by the modulation process itself. It is important to differentiate between the noise caused by the modulation process which is termed modulation noise and noise in the communication channel (thermal or distortion for example) which will be termed external or channel noise. The following sections deal with the modulation noise and the effect of channel noise is considered for digital modulation as a whole in Sections 6.6 and 6.7.

6.4.2. Modulation Noise from Quantising and Clipping (No Companding)

The way in which quantising noise arises has been explained in the previous section and this is subclassified as quantising noise which occurs within the maximum and minimum permitted levels and noise which occurs because the signal lies anywhere outside these limits. The former is the quantising noise proper and the latter is clipping noise.

In order to demonstrate a technique in determining the parameters of a system the baseband or modulating signal will be taken as a random variable of Gaussian amplitude distribution and for mathematical convenience the signal will be considered to be quantised first and then sampled, (this makes no difference to the practical result).

Figure 6.9 shows a signal and the type of waveform the error signal, due to quantising, will have. The 'sawtooth nature' of this error signal leads to the result that the mean square error noise is given by $s^2/12$ where s is the step voltage between coding levels. Intuitively the spectrum of the noise will be wider than the signal spectrum and in fact detailed analysis[2] shows that for a system with 128 different quantising levels, for example, the noise bandwidth is some hundreds of times wider than the signal. Since however this signal and its background of wideband noise is to be sampled it follows from Section 6.2. that the total noise contribution within the signal bandwidth will be $s^2/12$ if the sampling rate is exactly twice the top baseband frequency.

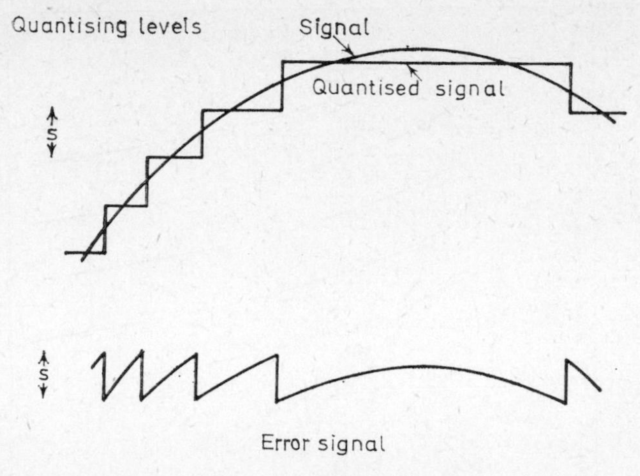

Figure 6.9 Signal quantisation

The noise arising from clipping is determined by considering the problem of a Gaussian noise waveform (which represents the speech signal) being transferred through a limiter characteristic. This limiter is such that the output $y(t)$ is related to the input $x(t)$ by the function $F(x)$ such that

$$F(x) = \begin{cases} -a & x < -a \\ x & |x| < a \\ +a & x > a \end{cases}$$

This problem is considered in detail in Chapter 5, Section 5.3.6 and the answer is quoted here as

$$f(a) = \frac{\operatorname{erf}^2 a/\sqrt{2}}{a^2 - \sqrt{(2/\pi)}a\, e^{-a^2/2} + (1-a^2)\operatorname{erf}(a/\sqrt{2}) - \operatorname{erf}^2(a\sqrt{/2})} \quad (6.1)$$

The total modulation noise consists of the sum of the quantising noise (N_Q) and the clipping noise (N_c) and it has been shown that the ratio of clipping noise to signal power is given by

$$\frac{N_c}{S_{\text{in}}} = f(a) \quad (6.2)$$

where S_{in} is the signal power, a is the ratio: clipping voltage/r.m.s. input voltage and $f(a)$ is given by equation 6.1 and is shown in Figure 6.10.

It has also been shown that,

$$N_Q = s^2/12$$

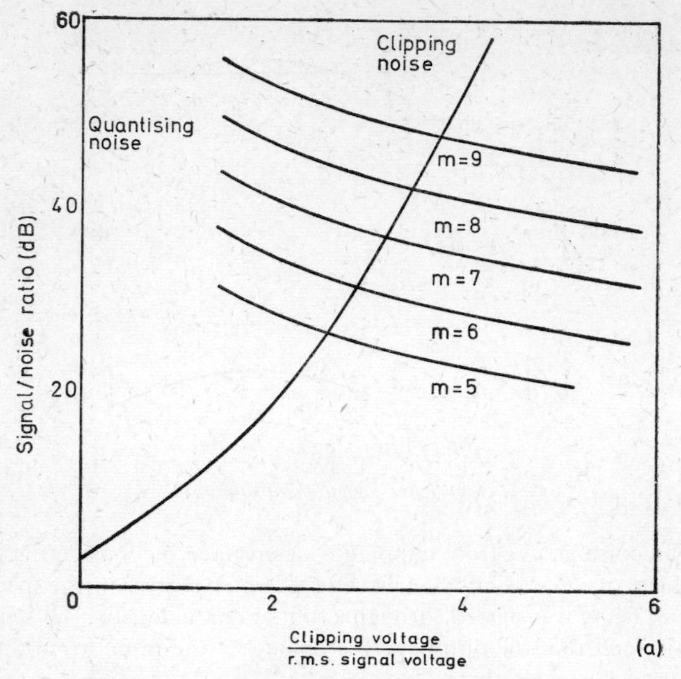

Figure 6.10 Quantising and clipping noise for p.c.m.

and

$$s = \frac{2\ (\text{clipping voltage})}{2^m}$$

where m = number of digits per sample.

Thus

$$\frac{N_Q}{S_{in}} = \frac{4}{12\times 2^{2m}}\,a^2 = \frac{a^2}{3\times 2^{2m}} \tag{6.3}$$

A plot of N_Q/S_{in} is plotted together with N_c/S_{in} in Figure 6.11 and the total modulation noise/signal ratio is given by

$$\frac{N}{S_{in}} = \frac{N_Q}{S_{in}} + \frac{N_c}{S_{in}} = \frac{a^2}{3\times 2^{2m}} + f(a) \tag{6.4}$$

Example 6.2

It is required to pulse code modulate an f.d.m. block of 300 channels each of 4 kHz bandwidth and average power −15 dBm0. The total noise in any channel is not to exceed 10 000 pW and the error noise under normal working conditions is to be negligible.

Calculate the number of digits per word required, the clipping voltage and the bit rate, assuming companding is not used.

Note that the expression dBm0 refers to dBs relative to 1 mW at a point of zero reference level (i.e. a point where the test signal tone would be 1 mW).

Solution

Noise power in a single channel $= 10\,000$ pW $= 10^{-8}$ W
$= -50$ dBmO

S/N ratio required $= -15+50 = 35$ dB

Turning now to Figure 6.10, it is necessary to select a value for m which will give a S/N of 35 and it can be seen that this requirement is just met for a value $m = 7$ for $a = \dfrac{\text{clipping voltage}}{\text{r.m.s. input voltage}}$ of about 3.6.

Thus the number of digits per word = 7

The total signal power at a point of zero reference level is $-15+$ $+10 \log 300$ dBm0.

i.e. Signal level $= -15+24{\cdot}7$
$= 9{\cdot}7$ dBm0
$= 9{\cdot}3$ mW.

If the impedance at this point is 600 Ω then the r.m.s. voltage is given by $\sqrt{(600\times 9{\cdot}3\times 10^{-3})} = 2{\cdot}36$ V.

Since a is equal to about 3·6, the clipping voltage is given by

Clipping Voltage $= 2{\cdot}36\times 3{\cdot}6 = 8{\cdot}5$ V.

The maximum modulating frequency is given by the product of the number of channels and their bandwidth i.e.

$f_m = 300\times 4 = 1200$ kHz

The sampling frequency is therefore given by twice this figure
Sampling frequency $= 2400\times 10^3$ samples per second
Since there are 7 bits per word the bit rate becomes

Bit rate $= 7\times 2400\times 10^3$ bits per second
$= 16\,800\times 10^3$
Bit rate $=$ *16·8 Mbits per second*

6.4.3. Modulation Noise from Quantising (With Companding)

In the previous section the signal to be encoded was sampled between limits in a uniform distribution of sampling amplitudes. This is a useful introduction to encoding and quantising noise analysis. However no practical p.c.m. system is likely to work with a uniform distribution of sampling amplitudes expecially when the waveform to be sampled is speech. Not only does speech have a highly nonlinear amplitude distribution, but the speech level varies from talker to talker. The level also depends on whether the telephone is close to or far from the exchange at which the signal is to be sampled and transmitted onwards as a p.c.m. pulse stream.

It is necessary to consider first how a nonlinear distribution of sampling amplitudes is an advantage and this can be seen by considering one particular talker power level. Most of the time the instantaneous amplitude of the speech waveform will be close to the mean level. If the decision amplitudes are closer together at levels close to the mean the quantisation noise at these levels will be reduced. Thus the objective, if only one speech level were involved, would be to ensure that the product of the quantisation noise for a given step in decision amplitude with the average amount of time the waveform spends between these two levels is kept constant. This means that the best nonlinear distribution of decision amplitudes for a waveform depends on the probability density function.

In practical p.c.m. systems for telephony, this system is called compression, and is used at the encoder. The corresponding complementary arrangement at the decoder is called expansion and the combination of both in a p.c.m. link is termed companding.

Before proceeding to establish the criteria required for a practical companding characteristic, it is worthwhile considering the method of calculating the quantising noise contribution in a system using companding. As in the previous section, the quantising noise will be sawtooth in nature, except that the 'teeth' as shown in Figure 6.9 will be of different amplitudes according to the voltage step between decision amplitudes. If the decision amplitudes are V_i, where $i = \pm 1, \pm 2, \pm 3 \ldots \pm c$, and where V_c and $V_{-c}(=-V_c)$ are the clipping voltages, then the quantisation power due to the lowest level will be $(V_1-0)^2/12$ multiplied by the proportion of time that the waveform spends between 0 and V_1. Thus the total quantising noise is,

$$N_Q = \sum_{n=-c}^{+c} \frac{(V_n-V_{n-1})^2}{12} [P(V_n)-P(V_{n-1})]$$

where $P(V_n)$ is the probability distribution of the waveform defined by

$$P(V_n) = \int_{-\infty}^{V_n} p(x)\,dx$$

If the V_i are known and if the probability density function of the waveform is known, then it is not difficult to obtain the value of N_Q using a computer. Note that in the special case of no companding,

$$(V_n - V_{n-1}) = (V_{n-1} - V_{n-2}) = s$$

and so,

$$N_Q = \frac{s^2}{12}[P(V_c) - P(V_{c-1}) + P(V_{c-1}) - P(V_{c-2}) \ldots P(V_{-c+1}) - P(V_{-c})]$$

i.e.

$$N_Q = \frac{s^2}{12}$$

The curve of signal/noise ratio due to quantising and clipping noise for $m = 7$ for no companding in Figure 6.11 was obtained directly from the analysis of Section 6.4.2 and Figure 6.10. From this curve it can be seen that for a signal/noise ratio of 25 dB (for example) the range of input levels permissible is about 14 dB. As indicated earlier, practical systems can have very wide ranges of input depending on the mean level of various talkers under different conditions, and in Reference 1 it is shown that ranges of the order of 35 dB are encountered in practice. In Reference 1 it is also indicated that the following continuous compression laws provide a useful basis for theoretical calculations.

$$\frac{V_{out}}{V_c} = \frac{A}{1 + \log_e A}\left(\frac{V_{in}}{V_c}\right) \qquad 0 \leqslant V_{in} \leqslant V_c/A$$

$$\frac{V_{out}}{V_c} = \frac{1 + \log_e (AV_{in}/V_c)}{1 + \log_e A} \qquad V_c/A \leqslant V \leqslant V_c$$

A is a constant which defines the signal level at which the companding law changes from a linear to a logarithmic law.

Figure 6.11 also shows the effect that this compression law has upon the signal/noise ratio due to quantising and clipping for the value $A = 100$ and $m = 7$ and it can be seen that the input range is greatly extended to include the required range of input signals.

In practice, continuous compression laws are not convenient and it is common to segment the compression law into 7 or 13 segments, the 13 segment version being shown in Figure 6.12.

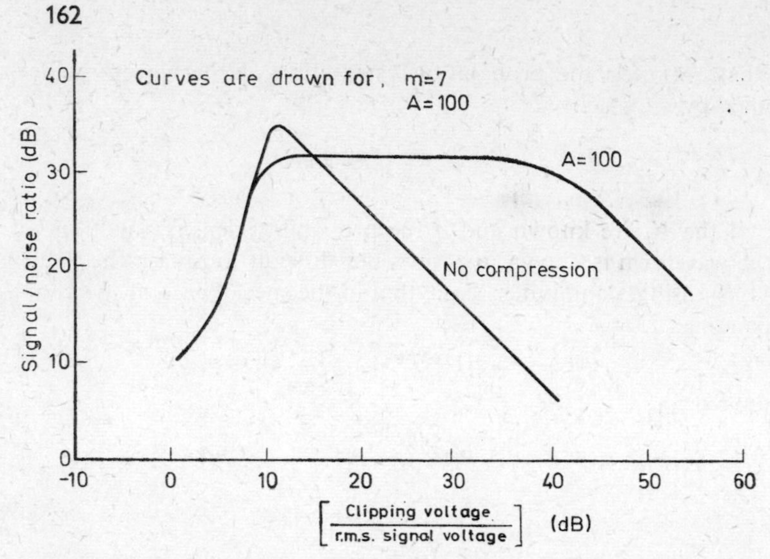

Figure 6.11 Signal/noise ratio due to quantising for signals with and without compression. A = 100 curve reproduced from Richards[1] by courtesy of the I.E.E.

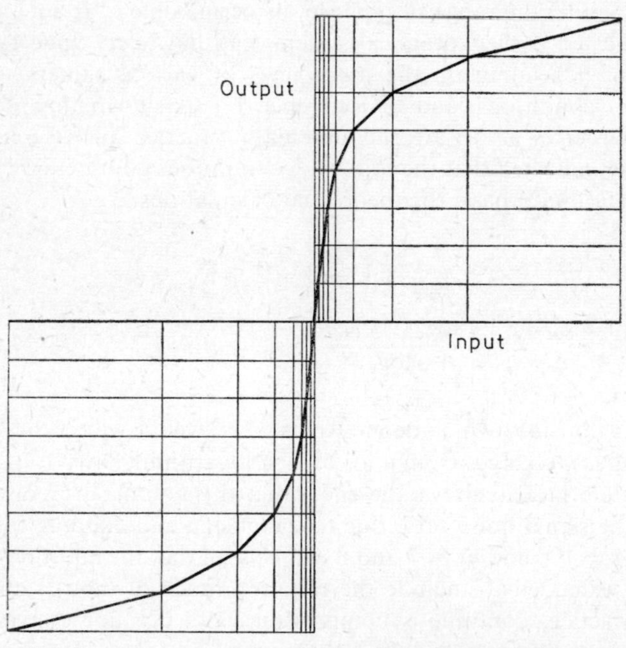

Figure 6.12 13 segment compandor characteristic. Reproduced from Richards[1] by courtesy of the I.E.E.

and corresponds to $A = 87 \cdot 6$. The signal/noise ratio due to quantising and clipping for the 13 segment characteristic is shown as the curve for zero errors in Figure 6.14.

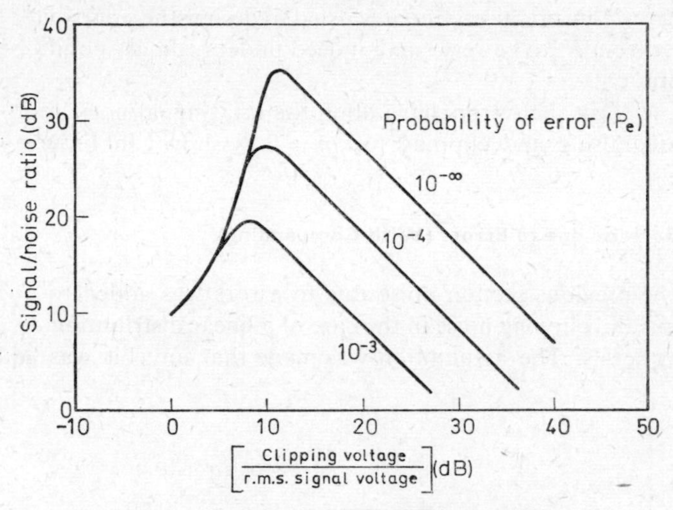

Figure 6.13 Noise due to errors (no companding)

6.4.4. Noise due to Errors (No Companding)

In Sections 6.6 and 6.7 it will be shown that noise in the communication channel gives rise to the possibility of the receiver making an error in the state of any binary pulse and thus there is a certain probability (usually small) that an error will be made. This probability is termed the probability of error P_e. The amount of noise the errors cause in the baseband depends upon which pulse or bit in the code word is corrupted. For example if the signal ranges between the clipping voltages $-V_c$ to $+V_c$ then if the most significant bit in the word is corrupted the amplitude is changed by V_c, if the second most significant bit is corrupted the error is $\pm \frac{1}{2} V_c$, etc. If it is assumed that any bit in the code word is equally likely to be corrupted then the average noise (N_e) due to these errors is given by

$$N_e = P_e V_c^2 [1 + \tfrac{1}{2} + \ldots (\tfrac{1}{4})^{2m-2}] \qquad (6.5)$$

which for a reasonably large value of m reduces to

$$N_e = \frac{P_e V_c^2 4}{3}$$

i.e. $$\frac{N_e}{S} = \frac{4P_eV_c^2}{3\times\text{signal power}} = \frac{4P_ea^2}{3} \qquad (6.6)$$

The dependency of N_e upon a^2 does not alter the optimum clipping level in practice since it is usual to design the contribution of noise from N_e to be very small indeed under ordinary conditions of operation.

The effect that errors have upon the total signal/noise ratio due to quantising and clipping for $m = 7$ is shown in Figure 6.13.

6.4.5. Noise due to Errors (With Companding)

In the previous section noise due to errors was added to quantisation and clipping noise in the case of a linear distribution of sampling levels. The assumption was made that any bit was equally

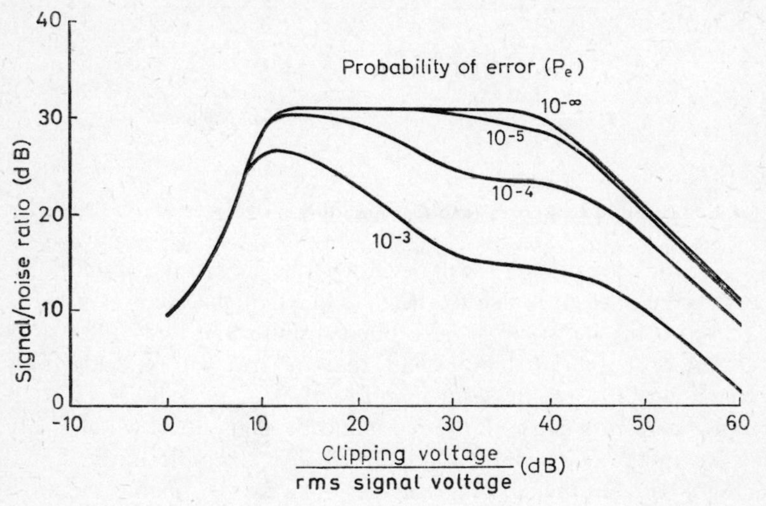

Figure 6.14 Noise due to errors (with companding). Curves are drawn for 13 segment compression law (equivalent to A-law, A = 87·6) m = 7

likely to be detected erroneously. Making the same assumption, it is possible to weight the error noise caused by each decision error according to the companding law, due to the fact that this determines the probability of an error occurring in a given compander segment. Using a digital computer this error noise can be added to the companded quantisation noise for any error rate. This information, obtained from Reference 13 is presented as Figure 6.14.

6.5. DELTA MODULATION AND SIGMA-DELTA MODULATION

6.5.1. Description of Modulation Process

The delta modulator uses the information in an analogue signal to code a stream of pulses at its output. Like conventional pulse code modulation, the pulses are equally spaced and of equal width and amplitude and the information is conveyed by the existence or

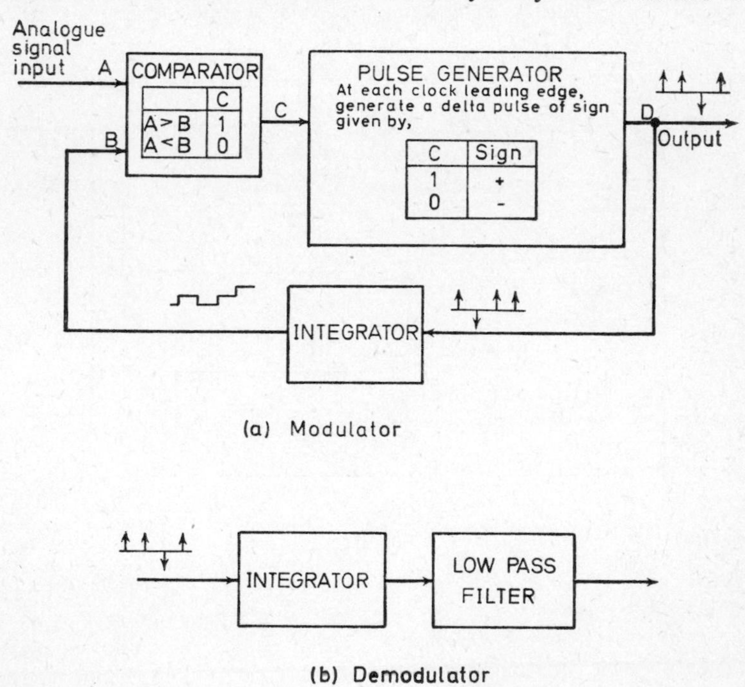

Figure 6.15 The delta modulator. (a) Modulator; (b) demodulator

non-existence of individual pulses. A block diagram of an ideal delta modulator is shown in Figure 6.15 where it can be seen that the output consists of either positive or negative going delta pulses. When the pulses are passed through an ideal integrator a staircase type of waveform is produced, a step upwards or downwards (depending on polarity) being produced for every pulse. The action of the delta modulator is to compare the height of such a generated waveform with the height of the analogue input waveform. At the end of each clock pulse period the comparison is used to decide the polarity of the next delta pulse. Thus the stair-

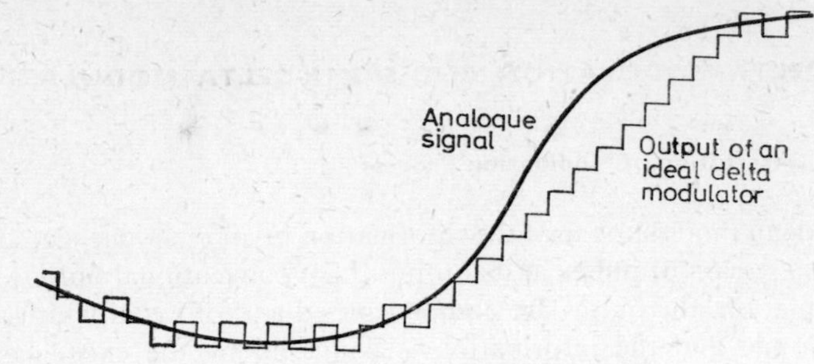

Figure 6.16 Quantising noise in delta modulation

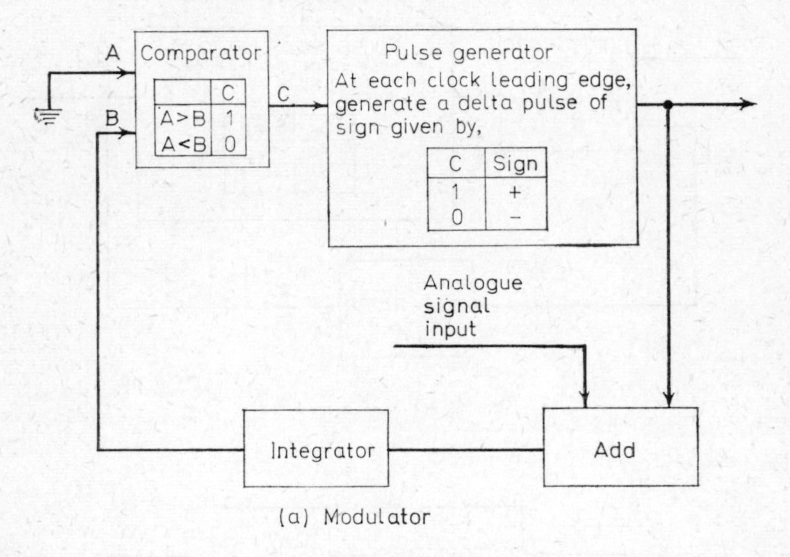

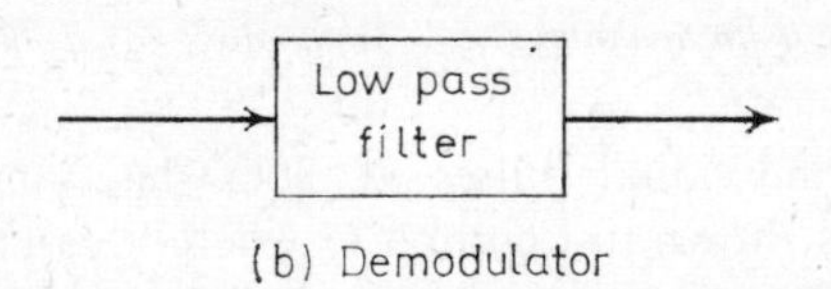

Figure 6.17 The sigma-delta modulator. (a) Modulator; (b) demodulator

case waveform produced by integration of the delta pulses follows the analogue signal in the quantised manner shown in Figure 6.16. The demodulation, therefore, consists simply of an integrator followed by a low pass filter.

A distortion other than quantising noise can arise as shown in Figure 6.16. If the analogue signal rises rapidly, the comparator may indicate that positive going delta pulses should be sent continuously over a period of time. This corresponds to a maximum slope of the quantised signal fed back to the comparator. If the analogue signal rises more rapidly than this slope, distortion will occur. Thus delta modulation limits not only the amplitude of the signal but the maximum rate of change of the signal. In a sine wave the maximum slope is proportional to amplitude and frequency, and so to avoid slope limiting, the maximum level of the input must decrease by 6 dB per octave increase in input frequency. This type of pre-emphasis is conveniently achieved by passing the input through an ideal integrator before modulation and omitting the integrator in the demodulator. The modulation process is now termed sigma-delta modulation and block diagrams of the modulator and demodulator are shown in Figure 6.17.

6.5.2. Quantising and Clipping Noise [5]

The noise performance of a sigma-delta modulator is analysed in this section. In the first instance, however, it is convenient to consider in more detail the process of delta modulation. This is then extended to sigma-delta modulation.

In order to separate the effects of quantising and clipping noise in delta modulation it will be assumed that the slope of the signal at the input to the delta modulator does not exceed the maximum permitted slope. In Figure 6.16, if the clock frequency is f_c and the height of each step is s then the maximum slope is $s.f_c$. The quantising noise is now given by the difference between the analogue signal and the staircase waveform. Since the decision of whether to send a positive or a negative pulse is taken at every clock pulse, the delta modulator can be considered to sample the analogue waveform at a rate of f_c and the staircase waveform of Figure 6.16 then results from the prolonging of the sample amplitudes for the whole clock period. The samples alone are shown in Figure 6.18a where it will be noticed that they are quantised in amplitude, and that they must change one unit step s (and only one) either positively or negatively every clock pulse.

It is now convenient to consider the samples as being formed by sampling a continuous quantised waveform as was done for the p.c.m. case. It turns out that this quantisation waveform is formed by sloped quantising levels of gradient sf_c, spaced vertically at intervals of $2s$. The continuous quantised waveform must remain on any quantisation slope for a minimum period of one clock pulse, a condition that is complied with if no slope limiting occurs.

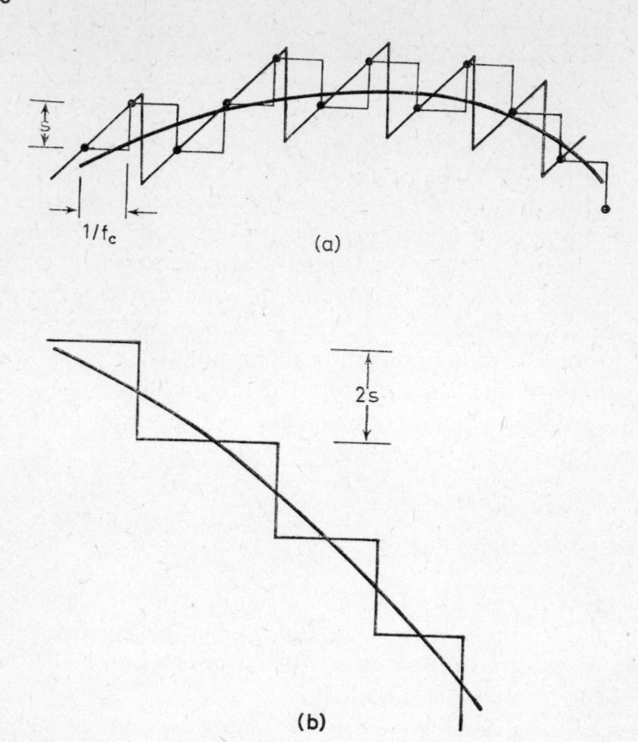

Figure 6.18 Sampling and quantising in delta modulation. (a) Sampling; (b) quantising of the signal minus ramp

If now a ramp of slope sf_c is subtracted both from the signal and the continuous quantized waveform the result of Figure 6.18b is obtained, where it can be seen that the new signal has been quantised in the same manner as in the p.c.m. case with a quantising step of $2s$.

The quantising noise in delta modulation can be obtained by following the arguments presented for pulse code modulation in Section 6.4.2, and the total quantising noise is initially taken as $(2s)^2/12$. As explained in Section 6.4.2 this total noise can be considered to be evenly spread between zero frequency and the clock frequency (f_c) and so the spectral density of the noise becomes,

$$N_0(f) = \frac{s^2}{3f_c} \tag{6.7}$$

This assumption is only true if the spectrum of the original quantising noise before sampling is reasonably level over the width of the clock frequency, which in view of the high clock frequencies used in delta modulation may not be strictly true.

Another reason why equation 6.7 is not strictly true arises from the fact that the signal under consideration does not have a Gaussian amplitude distribution but rather the distribution of a ramp plus random variable. Taken by itself the ramp would quantise to give 'noise' of a pure sawtooth waveform of frequency half that of the clock and the power spectrum of such a signal is anything but level. This lines up with the fact that under conditions of zero input a delta modulator changes from one state to the other every clock period. However under conditions of normal signal loading it is assumed that equation 6.7 applies, so it must be reckoned that the final calculated signal/noise ratio obtained may be a little pessimistic.

So far only delta modulation has been considered. However, these results can easily be extended to sigma-delta modulation because a sigma-delta modulator can be considered as a delta modulator with the signal integrated before input and the output differentiated. The quantising noise is therefore effectively differentiated in the sigma-delta modulator and the spectrum of the noise in this case becomes

$$N_0(f) = \frac{s^2(2\pi f)^2}{3f_c} \tag{6.8}$$

The output pulses of a sigma-delta modulator are frequently prolonged throughout the pulse period (non-return to zero pulses) and provided the decision mechanism in the sigma-delta modulator makes its decision at the end of the clock period, the step size s is given by the energy of the output pulse.

$$\text{i.e. } s = V_c/f_c \tag{6.9}$$

where V_c is the magnitude of either a positive or negative going pulse at the output (which is integrated and fed back to the input, as in the diagram in Figure 6.17) and f_c is the clock frequency.

Substituting in equation 6.8

$$N_0(f) = \frac{(2\pi f V_c)^2}{3f_c^3} \tag{6.10}$$

If the mean square value of the signal is v^2 and the signal has a flat spectrum extending from zero frequency to f_m, then the double sided power spectrum of the signal has a spectral density

$$S(f) = v^2/2f_m$$

and the signal-to-quantising noise ratio in a narrow frequency band at the frequency f is given by

$$\left(\frac{S(f)}{N_0(f)}\right)_Q = \frac{3f_c^3}{8\pi^2 f_m f^2}\left(\frac{v}{V_c}\right)^2 \tag{6.11}$$

Note that from equation 6.10 the total noise appearing in the bandwidth $-f_m$ to $+f_m$ which is obtained by integrating this equation with respect to f is given by

$$N_Q = \frac{8\pi^2 f_m^3 V_c^2}{9f_c^3}$$

The total signal/noise ratio at the output is therefore

$$\frac{S_{out}}{N_Q} = \frac{9f_c^3}{8\pi^2 f_m^3}\left(\frac{v}{V_c}\right)^2$$

Since the delta modulator limits the maximum slope of the signal, the sigma-delta modulator limits the amplitude and acts as a limiter of maximum amplitude V_c. The clipping mechanism is exactly the same as for the p.c.m. case except that in general the clock frequency is very much higher than twice the maximum modulation frequency. According to Section 6.2 therefore the noise due to clipping should decrease as the ratio f_c/f_m becomes larger, the exact decrease depending on the shape of spectrum. To avoid detailed analysis the graph of clipping noise against the ratio of clipping level to r.m.s. signal level for $f_c/f_m = 2$ may be used while recognising that the result may again be a little pessimistic.

6.5.3. Noise due to Errors [6]

In sigma-delta modulation the demodulator is simply a low pass filter and hence errors in pulses caused by channel noise will be simply added to the signal itself. If a pulse is changed from positive to negative or vice versa the mean square error noise added to the signal is $(2V_c)^2$ and if the probability of error is P_e there will be a total of about P_eN errors in a stream of N pulses (provided N is large). If the N pulses take a time w seconds to be transmitted the average noise power due to errors is

$$N_e = \frac{1}{w}\,\frac{P_eN(2V_c)^2}{f_c} = 4P_eV_c^2 \tag{6.12}$$

These errors occur in time with the clock pulses and so can be treated in effect as additional quantising noise. Thus the error noise is assumed to be evenly distributed over the frequency band equal to the clock frequency and the spectral density of noise due to errors becomes

$$\frac{4P_eV_c^2}{f_c}$$

The signal/noise ratio due to errors in a narrow bandwidth centred on a frequency in the signal band is thus given by

$$\left(\frac{S(f)}{N_0(f)}\right)_e = \frac{f_c}{8f_m}\left(\frac{v}{V_c}\right)^2 \frac{1}{P_e} \tag{6.13}$$

The total signal/noise due to errors is

$$\frac{S_{\text{out}}}{N_e} = \frac{f_c}{8f_m}\left(\frac{v}{V_c}\right)^2 \frac{1}{P_e}$$

6.6. SIMPLE DETECTION AND SECONDARY MODULATION

The output from the simplest of digital modulators consists of a stream of binary pulses. This is a stream where the pulses have two possible states, a positive pulse for one state and a negative pulse for the other for example. In the simplest of transmission systems, these pulses are transmitted directly over lines. As in any communication channel, the transmission line will have thermal noise added to it and possibly interference. The problem at the receiver therefore is to decide for each pulse interval, in a background of noise, which type of pulse out of two types is being sent. It is then possible for the receiver to decode the pulse stream, or what is more important, the receiver may now regenerate the original signal free of noise for retransmission. Thus although a repeater in a digital modulation link may make an error now and then, the signal is regenerated free of noise. This of course is in contrast to repeaters in analogue modulation links where noise is merely amplified and retransmitted.

Assume that a digital modulator produces a stream of pulses of either $+V$ volts or $-V$ volts, a possible method of detection might be to sample the pulse stream at approximately the middle of the pulse period and test whether this sample is greater or less than zero. Figure 6.19 shows the amplitude of the two possible pulses with a Gaussian distribution of noise added to each. If a sample taken in the middle of the pulse period is positive then it is obviously much more likely that the pulse sent was $+V$. If a sample taken in the middle of a pulse period is negative then the pulse is likely to

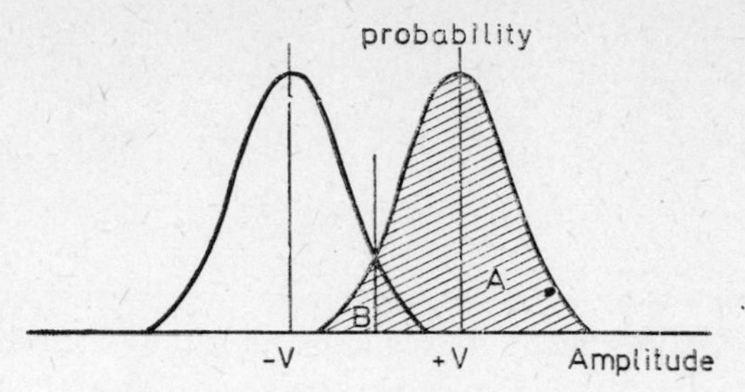

Figure 6.19 Simple detection of binary pulses

be $-V$. This therefore is a workable scheme for making a decision in this type of system.

If a positive pulse is being transmitted the probability of the sample being greater than zero is the area shaded A in Figure 6.19 while the probability of the sample being less than zero is the small remaining area B.

The probability of error in this case is therefore given by

$$P_e = \int_{-\infty}^{-V} \frac{1}{\sqrt{(2\pi N)}} e^{-\frac{x^2}{2N}} \, dx$$

where N is the total mean square value of the noise.

This method of detection therefore implies that careful filtering of the background noise could be used to reduce the total noise power to minimum. If the filter is too narrow however the signal pulse will be distorted but since the detector is only required to determine the presence or otherwise of a particular pulse and not its shape, this does not matter. Thus quite different forms of filtering can be used in digital systems and the concept of optimum pulse shape distortion and optimum sampling time for the output arises.

In the detection method proposed above (viz. sampling in the middle of the pulse) for instance, the signal over one period is a d.c. level and the noise is an a.c. type of signal. If therefore the total of signal plus noise is passed through an integrating device, the noise component will tend to cancel out and the d.c. level of the signal will tend to accumulate. Thus at the end of the pulse period the maximum advantage has been taken of the d.c. signal while the noise has been cancelled to a minimum.

Thus a better way of making a decision may be to first perform the calculation

$$\int_0^T y(t) \, dt \tag{6.14}$$

where T is the signal period and $y(t)$ is received signal and background noise. The decision is then based on whether the result of this calculation is positive or negative.

In radio systems, however, it is convenient to transmit a pulse train from a digital modulator by using the pulse to affect a continuous carrier in some way. Thus there is a double modulation process, the primary being the original analogue to digital modulation and the secondary being the modulation of the continuous carrier by the pulses. For example, a binary pulse train can be used to shift the frequency of a carrier from one frequency to another, or to reverse the phase of a carrier. In chirp modulation one state is a rising frequency and the other state a decreasing frequency.

If secondary modulation is used, the detection process of equation 6.14 will not work since the a.c. signal component will tend to be cancelled out as well as the noise. If, however, the incoming noisy signal is first multiplied by an exact replica of the transmitted signal, there will be an accumulation of the signal component as before when the product is integrated over the period of the signal. The detection method therefore consists of multiplying the incoming noisy signal separately by both of the possibilities of transmitted signal, integrating over the signal period, and choosing the signal which gives the highest output for the calculation process. This may be represented by,

$$\left.\begin{array}{l} \int_0^T y(t)\, S_1(t)\, \mathrm{d}t \\ \int_0^T y(t)\, S_2(t)\, \mathrm{d}t \end{array}\right\} \qquad (6.15)$$

Choose $S_1(t)$ or $S_2(t)$ as transmitted signal according to which of the calculations gives the largest result.

Here $S_1(t)$ and $S_2(t)$ are the possibilities for the transmitted signal. In fact this turns out to be the optimum detection process under certain conditions as will be shown in the next section. The integrals of equation 6.15 may be regarded as convolution integrals and the process of detection as a filtering process. This is further explained in Appendix 1. The impulse function of the filter is given by

$$h(t) = S(T-t)$$

It can be shown that for the case of white Gaussian noise this is the impulse function of the optimum filter for the signal $S(t)$ and the filter is termed a matched filter[12].

6.7. COHERENT DETECTION AND ERROR PROBABILITY 6-11

In order to provide a little more insight into the principles of matched filter detection and coherent detection in general, the following simplified analysis is given which is intended to be an introduction to the more rigorous treatments provided in the literature of the references.

In one simple detection method first considered in Section 6.6, it was proposed that a single sample should be taken half way through the signal (or pulse) period thus wasting all the information available in the remaining signal period on either side of the sample. It is reasonable to suppose that if more samples were taken during the signal period then much more information could be obtained thus increasing the likelihood of the receiver making the correct decision regarding which of the signal possibilities is being sent.

In order to demonstrate that this is in fact the case it is proposed that the receiver is capable of taking a large number of samples of the incoming signal and background of noise. The analysis applies equally well to a pulse signal in a pulse stream (primary digital modulation) as it does to a signal made up of a carrier modulated in some way by the pulse stream or any other type of secondary modulation.

In binary systems, two possible signals are sent, these are written as $S_1(t)$ and $S_2(t)$. These may be a positive pulse for $S_1(t)$ and a negative pulse for $S_2(t)$ in a pulse train; or a carrier of one frequency for $S_1(t)$ and a carrier of another frequency for $S_2(t)$ in the case of a frequency keyed secondary modulation system. Tertiary systems have three possible signals, positive pulses, no pulses, and negative pulses for example in a pulse train and, further, the most general of digital systems can be considered to have m possible signals. These are termed m-ary systems and may have m different amplitude levels in a pulse train for example or m different frequencies of a carrier, or m different phases of a carrier, in the case of secondary modulation. The m signals will be represented by $S_1(t)$, $S_2(t)$... ... $S_m(t)$.

Suppose that the signal S_i is being sent, the first of the many samples y_1, during the period of this signal is therefore made up of the sum of the first sample of the signal, S_i, and the noise sample n_1. Suppose that r samples are taken, then

$$y_1 = S_{i_1} + n_1$$

$$y_2 = S_{i_2} + n_2$$

$$\text{up to} \quad y_r = S_{ir} + n_r$$

The principle of the method for one sample of Section 6.6 may now be applied to each of these samples. Thus if each of the known possibilities for the m signals is subtracted from the set of r received samples, then for one set the subtraction will leave the noise components $n_1 n_2 \ldots n_r$ only. The one signal which gives this condition indicates, of course, that this was the original signal sent. It is therefore necessary to know the probability of receiving certain amplitudes for these r samples of noise and it is assumed that the background noise being sampled is of Gaussian amplitude distribution. The probability of finding the amplitude of a sample at a particular value is also Gaussian and it is further assumed that successive samples are uncorrelated. In this case the joint probability of the samples being at any set of amplitudes is the product of the separate probabilities. From equations 1.6 and 1.8

$$p(n_1, n_2, \ldots, n_r) = \frac{\exp \sum_{p=1}^{r} -n_p^2/2N}{(2\pi N)^{r/2}} \tag{6.16}$$

where N is the mean square value of the noise samples.

If the receiver therefore subtracts each of the m signals in turn from the received samples and computes the expression

$$\frac{\exp - \sum_{p=1}^{r} (y_p - S_{ip})^2/2N}{(2\pi N)^{r/2}}$$

it is required to choose the value of i which gives the largest result.

Thus the value of i is chosen which gives a minimum value for the expression

$$\sum_{p=1}^{r} (y_p - S_{ip})^2 = \sum_{p=1}^{r} (y_p^2 - 2y_p S_{ip} + S_{ip}^2)$$

Assuming that all the m signals have equal energy then $\sum y_p^2$ and $\sum S_{ip}^2$ will not be affected by which of the m signals is sent and the decision reduces to choosing the value of i corresponding to the maximum value of $\sum_{p=1}^{r} y_p S_{ip}$.

If now the number of samples is allowed to tend to infinity, the receiver is required to compute the expression over the signal period T

$$y' = \int_0^T y(t)\, S_i(t)\, \mathrm{d}t \tag{6.17}$$

for each value of i, and to select the largest.

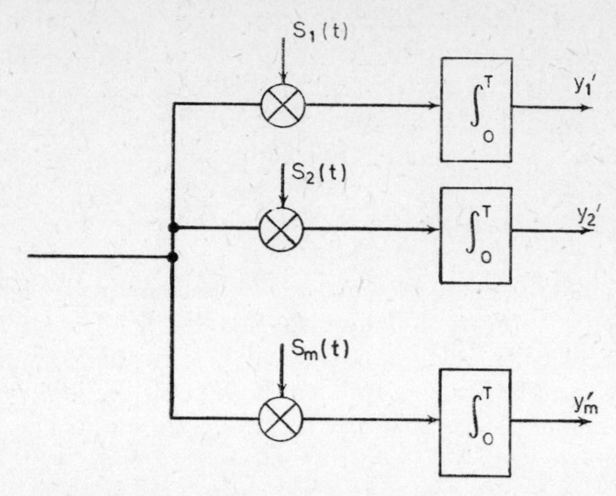

Figure 6.20 Correlation detector for an m-ary system

This corresponds to the result quoted in Section 6.6 for the matched filter, and the optimum detector may therefore be a set of filters or a set of correlators as shown in Figure 6.20.

The background noise causes the receiver to make the wrong decision now and again and the probability of error is calculated by considering the magnitudes of the outputs from the various correlators. Thus if the signal $S_j(t)$ is transmitted, the probability of correct decision (P_c) is the probability that the output from the S_jth correlator is greater than the outputs of all the other correlators.

If the signal being transmitted is $S_j(t)$ then the output of the S_ith correlator is given by equation 6.17 as

$$y' = \int_0^T y(t)\, S_i(t)\, \mathrm{d}t = \int_0^T [S_j(t)+n(t)]\, S_i(t)\, \mathrm{d}t \tag{6.18}$$

$n(t)$ is the noise component at the input and it can be seen that the noise component at the output is given by

$$Z_i = \int_0^T n(t)\, S_i(t)\, \mathrm{d}t \tag{6.19}$$

The error probability calculation uses the correlation between any two of these noise components and these are calculated as follows,

$$\overline{Z_j Z_k} = \overline{\int_0^T n(t)\, S_j(t)\, \mathrm{d}t \int_0^T n(t-\tau)\, S_k(t-\tau)\, \mathrm{d}(t-\tau)}$$

$$= \int_0^T \int_0^T R(\tau)\, S_j(t)\, S_k(t-\tau)\, \mathrm{d}(t-\tau) \tag{6.20}$$

Now it has already been assumed that the incoming noise to the system is white and for this particular case $R(\tau)$ is a delta function at $\tau = 0$. The integral of $R(\tau)$ gives the double sided spectral density N_0 of the noise and thus

$$\overline{Z_j Z_k} = N_0 \int_0^T S_j(t)\, S_k(t)\, \mathrm{d}t$$

$$= N_0 E\mu_{jk} \tag{6.21}$$

where E is the signal energy given by

$$E = \int_0^T [S_i(t)]^2\, \mathrm{d}t \tag{6.22}$$

and μ_{jk} is the correlation between signal S_j and signal S_k

i.e.
$$E\mu_{jk} = \int_0^T S_j(t)\, S_k(t)\, \mathrm{d}t \tag{6.23}$$

Now, if the signal S_1 (say) is transmitted, the probability of correct decision (P_c) is the probability that the output from the S_1 correlator $y_1' = E+Z_1$ is greater than the output of all the other correlators, $y_2' = E\mu_{12}+Z_2$, $y_3' = E\mu_{13}+Z_3 \ldots y_m' = E\mu_{1m}+Z_m$. This is the probability that Z_1 is at a certain value, combined with the joint probability that the other noise components $Z_k (k = 2$ to $m)$ which are not necessarily uncorrelated, are less than $E+Z_1-E\mu_{1k}$ integrated over all Z_1.

This may be written as

$$P_c = \int_{-\infty}^{\infty} \int_{-\infty}^{Z_1+E(1-\mu_{12})} \int_{-\infty}^{Z_1+E(1-\mu_{13})}$$

$$\ldots \int_{-\infty}^{Z_1+E(1-\mu_{1m})} \frac{\exp\left(-[1/(2N_0E)][\tilde{Z}][\mu]^{-1}[Z]\right)}{(2\pi N_0 E)^{m/2}\, |\mu|^{1/2}}\, \mathrm{d}Z_1\, \mathrm{d}Z_2 \ldots \mathrm{d}Z_m \tag{6.24}$$

The meaning of this equation is explained in the equations 1.36 and 1.37 of Chapter 1.

N_0 is the double sided noise spectral density, such that a filter of bandwidth B admits a noise power of $2N_0B$.

E is the energy of a signal bit, i.e. the power of the signal times the bit duration.

Example 6.3

Derive an expression for the error probability of a binary frequency shift keyed system in terms of the signal energy E and the noise spectral density N_0.

The two frequencies representing the two signals of the binary system will be uncorrelated and hence $\mu_{12} = 0$.

Substituting in equation 6.24 for $\mu_{12} = 0$

$$P_c = \int_{-\infty}^{\infty} \frac{\exp\,[-Z_1^2/(2N_0E)]}{(2\pi N_0 E)^{1/2}} \int_{-\infty}^{Z_1+E} \frac{\exp\,[-Z_2^2/(2N_0E)]}{(2\pi N_0 E)^{1/2}}\, dZ_1\, dZ_2 \quad (6.25)$$

This equation gives the answer but it may be tidied up in the following way, by first changing the variables to $\alpha = Z_1/\sqrt{E}$ and $\beta = Z_2/\sqrt{E}$

$$P_c = \int_{-\infty}^{\infty} \frac{\exp\,[-\alpha^2/2N_0]}{(2\pi N_0)^{1/2}} \int_{-\infty}^{\alpha+\sqrt{E}} \frac{\exp\,[-\beta^2/2N_0]}{(2\pi N_0)^{1/2}}\, d\beta\, d\alpha \quad (6.26)$$

This integral is tabulated in Reference 11 as a particular case of a more general set.

Equation 6.26 can be further simplified by treating it as a convolution integral. The characteristic function of the probability of correct decision is,

$$e^{N_0\psi^2/2} \cdot \frac{e^{-N_0\psi^2/2}}{j\psi} = \frac{e^{-N_0\psi^2}}{j\psi}$$

where ψ is the transform variable and so

$$P_c = \int_{-\infty}^{\sqrt{E}} \frac{\exp\,[-x^2/4N_0]}{(2\pi 2N_0)^{1/2}}\, dx \quad (6.27)$$

or

$$P_e = \int_{-\infty}^{-\sqrt{E}} \frac{\exp\,[-x^2/4N_0]}{(2\pi 2N_0)^{1/2}}\, dx \quad (6.28)$$

This may be written as

$$P_e = \tfrac{1}{2}\,[1 - \operatorname{erf} \sqrt{(E/4N_0)}] \quad (6.29)$$

The function

$$\tfrac{1}{2}\,(1 - \operatorname{erf} \sqrt{x})$$

is plotted in Figure 6.21.

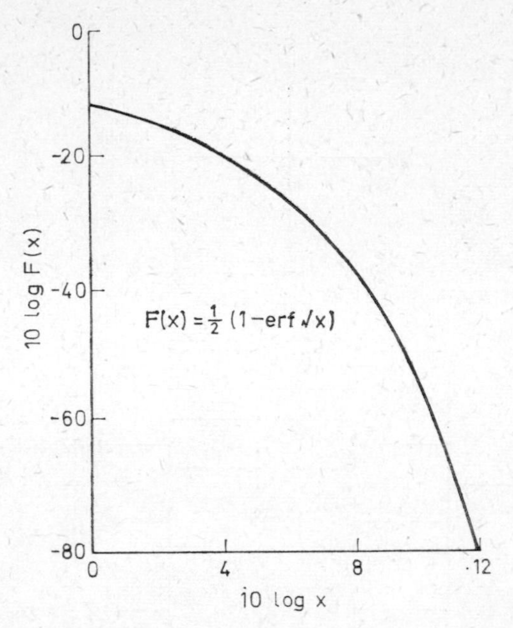

Figure 6.21 Error function in probability of error calculations

This equation shows rather more clearly than 6.24 that the probability of error is now dependent upon the signal energy i.e. signal × time and the noise spectral density rather than the total noise in the output of some filter.

Equation 6.24 must be modified slightly if there are pairs of signals which are the inverse of each other i.e. bi-orthogonal. This modification is illustrated in the following example.

Example 6.4

Derive an expression for the probability of error for the 4-phase bi-orthogonal system described below.

There are four signals S_1, S_2, S_3 and S_4 as shown in Figure 6.22a. S_1 is the inverse of S_2 and S_3 is the inverse of S_4, and the pair S_1 and S_3 are orthogonal to S_2 and S_4.

i.e.
$$\int_0^T S_1(t) S_3(t)\, dt = 0$$

Imagine for the sake of clarity that the correlation is performed for all four of the known signals (in fact correlation is only required for any two orthogonal signals since the decision with the inverse pair is merely a matter of sign).

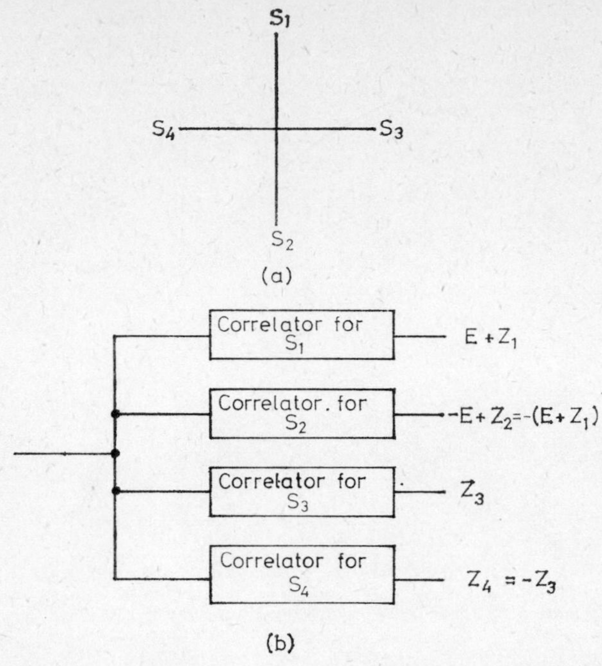

Figure 6.22(a) Representation of signals in bi-orthogonal coding; (b) correlation dectector for bi-orthogonal coding

If S_1 is sent for instance, the outputs of the correlators are as shown in Figure 6.22b and for correct decision it is required that:

$$E+Z_1 > -(E+Z_1) \quad \text{i.e.} \quad Z_1 > -E \qquad \text{EVENT 1}$$

$$\left.\begin{array}{l} E+Z_1 > Z_3 \\ E+Z_1 > -Z_3 \end{array}\right\} \quad \text{i.e.} \quad -(E+Z_1) < Z_3 < (E+Z_1) \quad \text{EVENT 2}$$

The probability of correct decision is now the joint probability of both events 1 and 2 occurring. Here Z_1 and Z_2 are uncorrelated so

$$P_c = \int_{E}^{\infty} \frac{\exp\,[-Z_1^2/(2N_0E)]}{(2\pi N_0 E)^{1/2}} \int_{-(Z_1+E)}^{Z_1+E} \frac{\exp\,[-Z_3^2/(2N_0E)]}{(2\pi N_0 E)^{1/2}}\, \mathrm{d}Z_1\, \mathrm{d}Z_3 \tag{6.30}$$

The method of making an optimum decision and calculating the error probability may be approached in a geometrical way as summarised in the following paragraphs.

In the previous analysis the signal and noise presented to the receiver were imagined to be sampled in time and then the number of samples allowed to tend to infinity. The alternative approach is to expand the signal plus noise in a series of orthogonal functions in a manner similar to Fourier analysis. If $n(t)$ is the noise component at the input to the receiver, then the expansion may be written as

$$n(t) = \sum_{j=1}^{N} Z_j \phi_j(t) \tag{6.31}$$

Here the $\phi_j(t)$ ($j = 1$ to N) are the normalised orthogonal functions (analogous to cos $n\omega t$ and sin $n\omega t$ in Fourier analysis) and the Z_j for a particular noise sample are a set of constants. Naturally $n(t)$ is different from one signal to the next and so the Z_j are in fact random variables.

The orthogonality property gives

$$\int_0^T \phi_j(t)\,\phi_k(t)\,\mathrm{d}t = \begin{cases} 1 & j = k \\ 0 & j \neq k \end{cases}$$
$$= \delta_{jk} \tag{6.32}$$

The random variables Z_j are given by

$$Z_j = \int_0^T n(t)\,\phi_j(t)\,\mathrm{d}t \tag{6.33}$$

The distribution of the Z_j are Gaussian if the distribution of $n(t)$ is Gaussian, and it may be shown[12] that the $\phi_j(t)$ are obtained from a solution of the equation

$$\int_0^T R(t-\tau)\,\phi_j(t)\,\mathrm{d}t = N_j \phi_j(\tau) \tag{6.34}$$

Where N_j is the variance of $n(t)$ and $R(\tau)$ is its autcorrelation function.

Considering the signal component in the input signal $y(t)$, a similar analysis is performed. Here it is required to represent the various possible signals $S_1(t)$, $S_2(t)$ ... $S_m(t)$, which may or may not be orthogonal to each other,
i.e.

$$S_1(t) = a_{11}\phi_1(t) + a_{12}\phi_2(t) + \;\ldots\; a_{1N}\phi_N(t)$$
$$S_2(t) = a_{21}\phi_1(t) + a_{22}\phi_2(t) + \;\ldots\; a_{2N}\phi_N(t)$$
$$\vdots$$
$$S_m(t) = a_{m1}\phi_1(t) + a_{m2}\phi_2(t) + \;\ldots\; a_{mN}\phi_N(t)$$

and $$a_{jk} = \int_0^T S_j(t)\, \phi_k(t)\, \mathrm{d}t \tag{6.34a}$$

The basis of this geometrical approach is to make the $\phi_j(t)$ axes in a space thus enabling the signal to be plotted as vectors in the space. For example in an m phase shift system, the axes could be cos ωt and sin ωt and the various angled signals then form vectors radiating from the origin in the normal way.

The next vital requirement is, of course, that the specific $\phi_j(t)$ for the signals should be the same $\phi_j(t)$ for the noise, i.e. the signals and the noise should have common axes. The important point is that this can only be so under very special circumstances, the requirement being the ability to expand the noise by any set of orthogonal functions depending on the nature of the signals. Now one condition for which equation 6.34 is satisfied for an arbitrary $\phi_j(t)$ is for $R(t)$ equal to a delta function at $t = 0$ i.e. the incoming noise is white and covering an infinite bandwidth.

Thus the signal can now be considered as being made up of the various components a_{ij} along the axes $\phi_j(t)$ and similarly the noise to be made up of the components Z_j (random variables) along the same axes $\phi_j(t)$.

The incoming signal plus noise $y(t)$ can thus be considered to be made up of components y_j' acting along the axes $\phi_j(t)$

i.e. $$y_j' = a_{ij} + Z_j \tag{6.35}$$

and from equations 6.33 and 6.34a,

$$y_j' = \int_0^T y(t)\, \phi_j(t)\, \mathrm{d}t \tag{6.36}$$

In order to obtain the components y_j' the receiver performs the operation of equation 6.36. Thus the receiver takes the form of a bank of correlators correlating the signal with the N $\phi_j(t)$ functions as shown in Figure 6.23.

Note that there are N correlators corresponding to the N orthogonal functions and that N can be less than m (e.g. 2 axes are required for m-phase working).

The decision process is arrived at in the same way as the previous approach except that the receiver is now required to select the value of i which yields the minimum value for the expression

$$\sum_{j=1}^{N} (y_j' - a_{ij})^2 \tag{6.37}$$

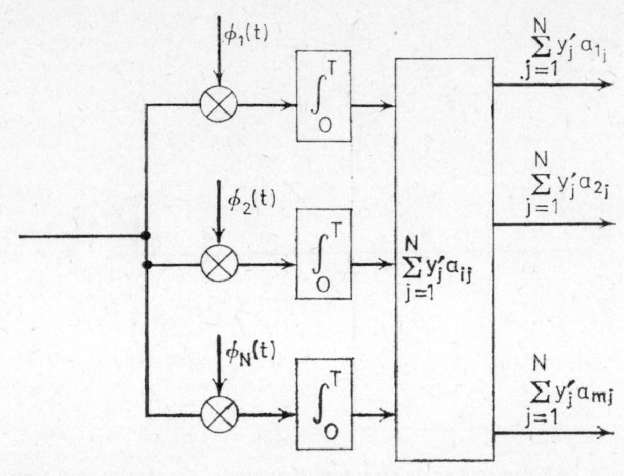

Figure 6.23 Alternative correlation detector for an m-ary system

As before this corresponds to choosing the largest value of expressions

$$\sum_{j=1}^{N} y_j' a_{ij} \tag{6.38}$$

Thus the correlators must be followed by a device which computes the expression of equation 6.38 as shown in Figure 6.23.

Error probabilities are calculated on a geometrical basis, taking the length of the signal vectors as $\sqrt{E}$ (E is the signal energy). The length of the noise components along the axes are calculated in exactly the same way as the noise components Z_j in the previous analysis except that $\phi(t)$ is substituted for $S(t)$. Thus, from 6.19 and 6.21

$$\overline{Z_j Z_k} = N_0 \mu_{jk}$$

However, since the $\phi_j(t)$ are all orthogonal $\mu_{jk} = \delta_{jk}$ and the length of each noise component is N_0.

Example 6.5

Derive the probability of error expression in Example 6.3 for a binary frequency shift keyed system.

In this system there are two signals which are already orthogonal (or uncorrelated) hence the geometrical space requires two axes ϕ_1 and ϕ_2 at right angles. (See Figure 6.24.) Signal S_1 lies along the one axis and is of length $\sqrt{E}$ signal S_2 lies along the other axis and is also of length $\sqrt{E}$. The problem is to decide that if S_1 was

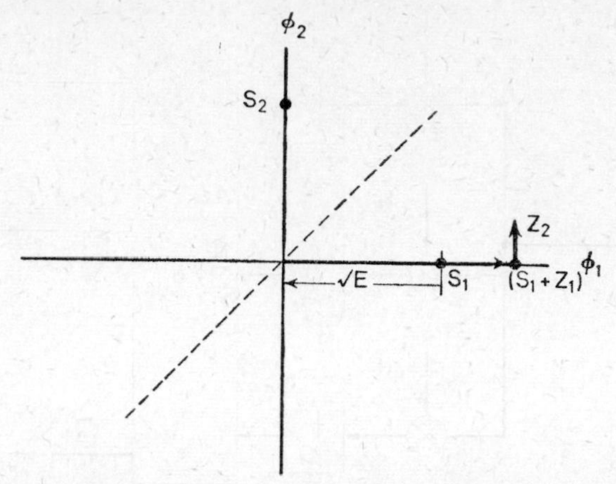

Figure 6.24 Geometrical representation of binary frequency shift keying

sent whether the received vector $(\mathbf{S}_1+\mathbf{Z}_1+\mathbf{Z}_2)$ lies on the $\mathbf{S}_1$ side of the decision boundary (dotted), or not. From Figure 6.24 it is seen that the probability of receiving S_1 correctly is the probability that Z_2 is less than the sum of the signal component and the noise component in the ϕ_1 direction, $\sqrt{E}+Z_1$. Thus the probability of correct decision P_c is the joint probability that Z_1 is at a certain value and that Z_2 is less than $\sqrt{E}+Z_1$ integrated over all Z_1. Remembering that the variance of Z_1 and Z_2 is N_0,

$$P_c = \int_{-\infty}^{\infty} \frac{e^{-Z_1^2/2N_0}}{(2\pi N_0)^{1/2}} \int_{\infty}^{Z_1+\sqrt{E}} \frac{e^{-Z_2^2/2N_0}}{(2\pi N_0)^{1/2}} dZ_1 . dZ_2 \qquad (6.39)$$

This is the same result obtained in equation 6.26.

It is interesting to rotate the axes of Figure 6.24 to obtain Figure 6.25 and now it can be seen that the probability of correct decision is the probability that the noise component Z_1 is less than $\sqrt{(E/2)}$

i.e.
$$P_t = \int_{-\infty}^{\sqrt{(E/2)}} \frac{e^{-Z_1^2/2N_0}}{(2\pi N_0)^{1/2}} dZ$$

Putting $Z_1 = x/\sqrt{2}$

$$P_c = \int_{-\infty}^{-\sqrt{E}} \frac{e^{-x^2/4N_0}}{(2\pi 2N_0)^{1/2}} dx$$

which is the same as equation 6.27.

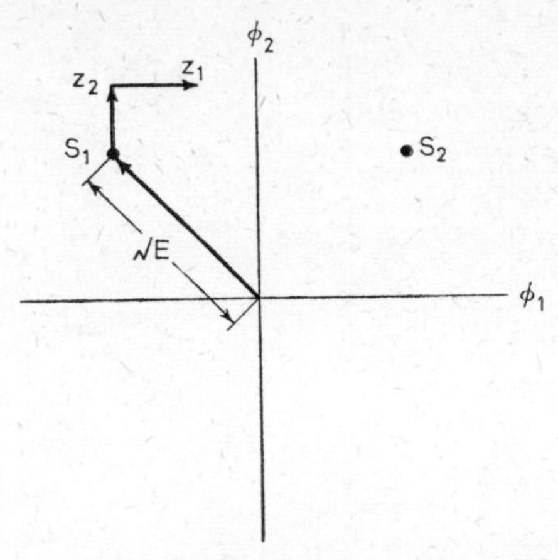

Figure 6.25 Alternative axes for binary frequency shift keying

Example 6.6

Derive the probability of error for a 4-phase bi-orthogonal system as in Example 6.4.

Here as for any number of phases only two axes are needed and the problem is shown in Figure 6.26.

It is seen that the probability of correct decision is that,

$$Z_1 > -\sqrt{E} \text{ and that } Z_3 < \sqrt{E}+Z_1$$
$$\text{and } Z_3 > -(\sqrt{E}+Z_1)$$

i.e.
$$P_c = \int_{-\sqrt{E}}^{\infty} \frac{e^{-Z_1^2/2N_0}}{(2\pi N_0)^{1/2}} \cdot \int_{-(\sqrt{E}+Z_1)}^{(\sqrt{E}+Z_1)} \frac{e^{-Z_3^2/2N_0}}{(2\pi N_0)^{1/2}} dZ_1\, dZ_3$$

Again a change in variable yields the same result as equation 6.30.

Example 6.7

An f.d.m. block of 300 channels of telephony each of 4 kHz bandwidth and average power −15 dBm0 (see example 6.2) is to be pulse code modulated and transmitted over a satellite link using binary phase reversal keying. Under fair weather conditions the total noise in any channel is not to exceed 10 000 pW. Under the worst possible conditions of reception the noise due to thermal

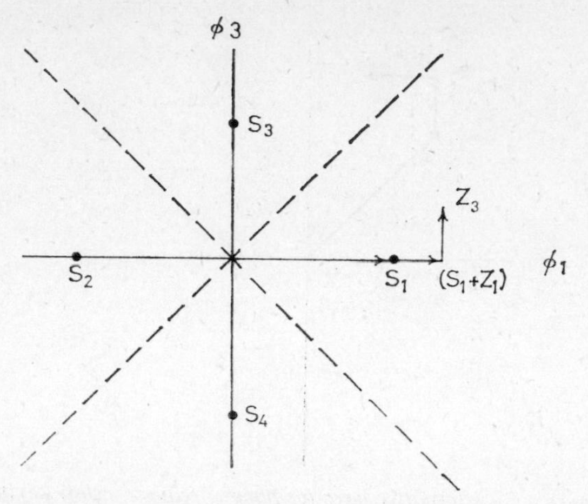

Figure 6.26 Geometrical representation of bi-orthogonal coding

effects is degraded by 6 dB and the total noise in any channel under these conditions is not to exceed 22 000 pW.

Calculate the ratio $10 \log \left[\dfrac{\text{Signal carrier power}}{\text{System noise temperature}}\right]$

required at the input to the receiver under normal working conditions.

Solution

Under fair weather conditions the noise in any channel is taken as being due only to the quantising and clipping caused by the modulation method itself. The allowable noise is 10 000 pW and the problem corresponds exactly to Example 6.2 where it is shown that the number of bits per code word required is 7.

Under unfavourable conditions of reception the noise allowable is 22 000 pW and so under these conditions the noise due to errors is taken as 12 000 pW.

i.e. Noise due to errors $= 12\,000$ pW $= 1{\cdot}2\times10^{-8}$ W
$= -49{\cdot}2$ dBm0 (dB relative to 1 mW)

Signal/noise ratio due to errors (S/N_e) $= -15+49{\cdot}2 = 34{\cdot}2$ dB

From equation 6.6 $\quad S/N_e = \dfrac{3}{4P_e a^2}$

From Example 6.2, the ratio
(clipping voltage/r.m.s. signal voltage) $a = 3{\cdot}6$
The probability of error P_e is given by

$$4{\cdot}7 - 6{\cdot}0 - 10 \log P_e - 11{\cdot}1 = 34{\cdot}2$$

$$10 \log P_e = -46{\cdot}6$$

From Example 6.3 $\quad P_e = \frac{1}{2}\,[1 - \text{erf}\,\sqrt{(E/2N_0)}]$

From Figure 6.21 $\quad E/2N_0 = 9{\cdot}3$ dB

Signal carrier power $\quad S = E/T$

where T is the duration of a signal bit.
The noise power spectrum is given by $\quad 2N_0 = kT_s$
where k is Boltzmann's Constant
T_s is the system noise temperature
and where it should be noted that the '2' arises because N_0 is the double sided power spectral density.

Thus
$$S/T_s = \frac{Ek}{2N_0 T}$$

from Example 6.2 the bit rate is found to be $1{\cdot}68 \times 10^7$

Thus
$$S/T_s = 9{\cdot}3 + 67{\cdot}7 - 228{\cdot}6 \text{ dB}$$
$$S/T_s = -151{\cdot}6 \text{ dB}$$

This is the signal/noise temperature under unfavourable conditions in which the noise is 6 dB higher than under fair weather conditions. Thus under normal working conditions the required ratio is

$$S/T_s = -145{\cdot}6 \text{ dB}$$

Problems

6.1 Calculate the bit rate required to transmit a signal of maximum frequency 10 kHz using p.c.m. if the number of bits per code word is 8. If the signal has a Gaussian amplitude distribution what is the maximum signal/noise ratio available?
(Ans. 160 kHz, 41 dB.)

6.2 A sigma delta modulator is modulated by a baseband of maximum frequency 108 kHz using a clock frequency of 1·7 MHz. Assuming the signal has a Gaussian amplitude distribution, calculate the maximum signal/noise ratio available in a narrow frequency slot centred on 74 kHz in the baseband. (Note. This problem may be solved by superimposing the result of equation 6.11 onto the clipping noise curve of equation 6.1 in exactly the same way as for p.c.m.)

(Ans. 17 dB)

6.3 A binary code is transmitted over a radio link using amplitude shift keying (i.e. 1 corresponds to carrier on, 0 corresponds to carrier off) at a bit rate of 1 Mbit/s. If the ratio of the carrier power (when keyed on) to noise temperature at the input to the demodulator is −157 dB, calculate the probability of error to be expected. (Note. Since the two possible signals are uncorrelated, the problem corresponds exactly to Example 6.3)

(Ans. $10 \log P_e = 46$)

6.4 Reformulate Example 6.4 to show that an expression for the probability of correct decision for binary phase reversal keying is

$$P_c = \int_{-E}^{\infty} \frac{\exp\,[-z^2/(2N_0E)]}{(2\pi N_0 E)^{1/2}}\, dz$$

Use the change in variable $\alpha = z/\sqrt{E}$ to show that the probability of error may be expressed as

$$P_e = \tfrac{1}{2}\,[1 - \operatorname{erf} \sqrt{(E/2N_0)}]$$

6.5 Use the geometrical approach to confirm the answer of Problem 6.4.

References

1. RICHARDS, D. L., 'Transmission performance of telephone networks containing p.c.m. links.' *Proc. I.E.E.* **115,** 1245 (1968)
2. BENNETT, W. R., 'Spectra of quantized signals' *B.S.T.J.*, **27,** 446 (1948)
3. LANNING, J. H. and BATTIN, R. H., *Random Processes in automatic control.* McGraw Hill Book Co., Inc. New York (1956)
4. VAN VLECK, H. J. and MIDDLETON, D., 'The spectrum of clipped noise' *Proc. I.E.E.E.*,**2,** 54 (1966)
5. INOSE, H., YASUDA, Y. and MURAKAMI, J., 'A telemetering system by code modulation delta-sign modulation.' *I.R.E. Trans. Space Electron. Telem.* **SET-8,** 204 (1962)
6. WOLF, J. K., 'Effects of channel errors on delta modulation.' *I.E.E.E. Trans. Commun. Technol.*, **Com-14,** 2 (1966)
7. ARTHURS, E. and DYM, H., 'On the optimum detection of digital signals in the presence of white Gaussian noise—a geometrical interpretation and a study of three basic transmission systems.' *I.R.E. Trans. on Commun. Syst.*, **CS-10** 336 (1962)

8. CAHN, C. R., 'Performance of digital phase-modulation communication systems.'
9. WOZENCRAFT, J. M., and JACOBS, I. M., *Principles of communication engineering*, John Wiley and Sons (1968)
10. NUTTALL, A. H., 'Error probabilities for equicorrelated *M*-ary signals under phase-coherent and phase incoherent reception,' *I.R.E. Trans. Inf. Theory*, **IT-8** 305 (1962)
11. GOLOMB, S. W. (Ed.) *Digital communications with space applications*, Prentice Hall, (1964)
12. DAVENPORT, W. B., and ROOT, W. L., *Random Signals and Noise*, McGraw Hill, New York (1958)
13. RICHMAN, G. D. 'Error noise with companding', *Private communication*, Post office Engineering Dept.

Appendix 1

THE FOURIER TRANSFORM AND ITS PROPERTIES

N.B. This topic is dealt with in a very comprehensive manner in Reference 4 of Chapter 1.

A1.1 DEFINITIONS

The Fourier transform used in this book is defined as

$$F(f) = \int_{-\infty}^{\infty} f(t)\, e^{-j\omega t}\, dt$$

$$f(t) = \int_{-\infty}^{\infty} F(f)\, e^{j\omega t}\, df$$

$$\text{where } \omega = 2\pi f$$

The following abreviation is used for 'transforms to'

$$F(f) \rightarrow f(t)$$
$$f(t) \rightarrow F(f)$$

A1.2 DICTIONARY OF FOURIER TRANSFORMS

$f(t)$	$F(f)$
$1 \quad \lvert t\rvert < \frac{1}{2}$ $0 \quad \lvert t\rvert > \frac{1}{2}$	$\dfrac{\sin \pi f}{\pi f}$
$\exp(-\pi t^2)$	$\exp(-\pi f^2)$
1	$\delta(f)$
$\cos \pi t$	$\frac{1}{2}[\delta(f-\frac{1}{2})+\delta(f+\frac{1}{2})]$
$\sin \pi t$	$j\,\frac{1}{2}[\delta(f+\frac{1}{2})-\delta(f-\frac{1}{2})]$

The Fourier transform of many functions arising in communication theory can be formed using the few basic functions given above in conjunction with the transform theorems that follow. Note that the symmetry of the Fourier transform allows the reverse transform of any function to be formed very easily.

A1.3 FOURIER TRANSFORM THEOREMS

1. Addition

If $f(t) \rightarrow F(f)$
and $g(t) \rightarrow G(f)$
then $f(t)+g(t) \rightarrow F(f)+G(f)$

2. Multiplication/Convolution

If $f(t) \rightarrow F(f)$
and $g(t) \rightarrow G(f)$
then $f(t) * g(t) \rightarrow F(f)G(f)$

$$f(t) * g(t) = \int f(u)g(t-u)\, \mathrm{d}u$$

3. Similarity

If $f(t) \rightarrow F(f)$
then $f(\alpha t) \rightarrow |\alpha|^{-1}F(f/\alpha)$

4. Shift

If $f(t) \rightarrow F(f)$
then $f(t-t_0) \rightarrow \mathrm{e}^{-\mathrm{j}\omega t_0}F(f)$
and $F(f-f_0) \rightarrow \mathrm{e}^{\mathrm{j}\omega_0 t}f(t)$
where $\omega_0 = 2\pi f_0$

5. Modulation

If $f(t) \rightarrow F(f)$
then $f(t) \cos \omega_0 t \rightarrow \frac{1}{2}[F(f-f_0)+F(f+f_0)]$
where $\omega_0 = 2\pi f_0$

6. Power

$$\text{If} \qquad f(t) \rightarrow F(f)$$

$$\text{then} \quad \int_{-\infty}^{\infty} |f(t)|^2 \, dt = \int_{-\infty}^{\infty} |F(f)|^2 \, df$$

7. Autocorrelation

If the autocorrelation function of $f(t)$ is $R(\tau)$ and if the power spectrum of $f(t)$ is $S(f)$

$$\text{then} \qquad R(\tau) \rightarrow S(f)$$

8. Derivatives

$$\text{If} \qquad f(t) \rightarrow F(f)$$

$$\text{then} \quad \frac{d}{dt} f(t) \rightarrow j\omega F(f)$$

$$\text{and} \quad \int f(t) \, dt \rightarrow \frac{1}{j\omega} F(f)$$

Appendix 2

DERIVATION OF MULTIPLE MOMENTS OF GAUSSIAN VARIABLES

Multiple moments of Gaussian variables may be expressed in second moments using the following formula, the derivation of which is demonstrated in this Appendix.

$$E(x_1.x_2.x_3 \ldots x_N) = \sum_{\text{all pairs}} \left[\prod_{\substack{j\neq k\\ j=1\\ k=1}}^{N} (x_j.x_k) \right] \qquad \text{(A2.1)}$$

The Gaussian variables may be defined by their multivariate characteristic function as follows (see equation 1.38)

$$M(\mathrm{j}v_1, \mathrm{j}v_2, \ldots, \mathrm{j}v_N) = \exp\left(-\tfrac{1}{2}\sum_{n=1}^{N}\sum_{m=1}^{N} \mu_{nm}v_n v_m\right) \qquad \text{(A2.2)}$$

or,

$$E\left[\exp\left(\mathrm{j}\sum_{n=1}^{N} v_n x_n\right)\right] = \exp\left(-\tfrac{1}{2}\sum_{n=1}^{N}\sum_{m=1}^{N} \mu_{nm}v_n v_m\right)$$

Substituting v_n for $\mathrm{j}v_n$,

$$E\left[\exp\sum_{n=1}^{N} v_n x_n\right] = \exp\left(\tfrac{1}{2}\sum_{n=1}^{N}\sum_{m=1}^{N} \mu_{nm}v_n v_m\right) \qquad \text{(A2.3)}$$

An expression for the multivariate characteristic function for any set of variables is useful because it provides an easy method of generating the multiple moments. For example if the left hand side of equation A2.3 is differentiated with respect to v_1 and $v_1, v_2 \ldots v_m$ are set to zero, the first moment of a variable in the set is formed. i.e.

$$\left.\frac{\partial}{\partial v_1}\right|_{v_1, v_2, \ldots v_N=0} E\left(\exp\sum_{n=1}^{N} v_n x_n\right) = E(x_1)$$

Similarly by differentiating with respect to all the variables, the left hand side of equation A2.1 can be formed.

$$\left.\frac{\partial^n}{\partial v_1\,\partial v_2 \ldots \partial v_N}\right|_{v_1,\,v_2,\,\ldots\,v_N=0} E\left(\exp \sum_{n=1}^{N} v_n x_n\right) = E(x_1.\,x_2.\,\ldots\,.x_N) \tag{A2.4}$$

Equation A2.4 shows therefore that the multiple moment of N Gaussian variables may be found by differentiating the right hand side of equation A2.3 with respect to $v_1, v_2 \ldots v_N$ and setting $v_1, v_2 \ldots v_N$ equal to zero.

For convenience put

$$\exp\left(\tfrac{1}{2}\sum_{n=1}^{N}\sum_{m=1}^{N} \mu_{nm} v_n v_m\right) = \exp(M)$$

Now,

$$\frac{\partial}{\partial v_k}\exp(M) = \exp(M)\frac{1}{2}\left[\sum_{m=1}^{N}\mu_{km}v_m + \sum_{n=1}^{N}\mu_{nk}v_n\right]$$

Again for convenience, this is abbreviated to

$$\frac{\partial}{\partial v_k}\exp(M) = \exp(M)\,[M_k]$$

In particular,

$$\frac{\partial}{\partial v_1}\exp(M) = \exp(M)\,[M_1]$$

$$\frac{\partial^2}{\partial v_1\,\partial v_2}\exp(M) = \exp(M)[M_2][M_1] + \exp(M)\tfrac{1}{2}[\mu_{12}+\mu_{21}] \tag{A2.5}$$

Putting $v_1 = v_2 \ldots v_N = 0$ and since $\mu_{12} = \mu_{21}$ A2.5 becomes

$$\left.\frac{\partial^2}{\partial v_1\,\partial v_2}\right|_{v_1,\,v_2,\,\ldots\,v_N=0} \exp(M) = E(x_1.x_2) = \mu_{12} \quad (\text{or } \mu_{21})$$

as would be expected.
Differentiating again

$$\begin{aligned}\frac{\partial^3}{\partial v_1\,\partial v_2\,\partial v_3}\exp(M) = {} & \exp(M)[M_3][M_2][M_1] \\ & + \exp(M)\tfrac{1}{2}[\mu_{23}+\mu_{32}][M_1] \\ & + \exp(M)[M_2]\tfrac{1}{2}[\mu_{13}+\mu_{31}] \\ & + \exp(M)[M_3]\tfrac{1}{2}[\mu_{12}+\mu_{21}]\end{aligned} \tag{A2.6}$$

Putting $v_1, v_2 \ldots v_N = 0$ each term becomes zero and so,

$$\left.\frac{\partial^3}{\partial v_1\,\partial v_2\,\partial v_3}\right|_{v_1,\,v_2,\,\ldots\,v_N=0} \exp(M) = E(x_1.x_2.x_3) = 0$$

Differentiating A2.6 again,

$$\begin{aligned}
\frac{\partial^4}{\partial v_1\,\partial v_2\,\partial v_3\,\partial v_4}\exp(M) = {} & \exp(M)[M_4][M_3][M_2][M_1] \\
& + \exp(M)\tfrac{1}{2}[\mu_{34}+\mu_{43}][M_2][M_1] \\
& + \exp(M)[M_3]\tfrac{1}{2}[\mu_{24}+\mu_{42}][M_1] \\
& + \exp(M)[M_3][M_2]\tfrac{1}{2}[\mu_{14}+\mu_{41}] \\
& + \exp(M)[M_4]\tfrac{1}{2}[\mu_{23}+\mu_{32}][M_1] \\
& + \exp(M)\tfrac{1}{2}[\mu_{23}+\mu_{32}]\tfrac{1}{2}[\mu_{14}+\mu_{41}] \\
& + \exp(M)[M_4][M_2]\tfrac{1}{2}[\mu_{13}+\mu_{31}] \\
& + \exp(M)\tfrac{1}{2}[\mu_{24}+\mu_{42}]\tfrac{1}{2}[\mu_{13}+\mu_{31}] \\
& + \exp(M)[M_4][M_3]\tfrac{1}{2}[\mu_{12}+\mu_{21}] \\
& + \exp(M)\tfrac{1}{2}[\mu_{34}+\mu_{43}]\tfrac{1}{2}[\mu_{12}+\mu_{21}]
\end{aligned} \tag{A2.7}$$

Putting $v_1, v_2, \ldots, v_N = 0$ and since pairs like $\mu_{nm} = \mu_{mn}$ equation A2.7 becomes,

$$\begin{aligned}
\left.\frac{\partial^4}{\partial v_1\,\partial v_2\,\partial v_3\,\partial v_4}\right|_{v_1,\,v_2,\,\ldots\,v_N=0} \exp(M) &= E(x_1.x_2.x_3.x_4) \\
&= \mu_{23}\mu_{14}+\mu_{24}\mu_{13}+\mu_{34}\mu_{12}
\end{aligned}$$

Continuation of the process will give the general result of equation A2.1.

USEFUL FORMULAE FOR FREQUENCY MODULATION

The results of this Appendix stem from obtaining the power spectrum of the function $V(t) = \sin [x_1(t)+x_2(t)]$, where $x_1(t)$ and $x_2(t)$ are general time varying functions.

Now $\quad V(t) = \sin x_1(t) \cos x_2(t)+\cos x_1(t) \sin x_2(t)$

So
$$\begin{aligned} R(t) = & E[\sin x_1(\tau) \cos x_2(\tau) \sin x_1(\tau+t) \cos x_2(\tau+t] \\ & +E[\cos x_1(\tau) \sin x_2(\tau) \cos x_1(\tau+t) \sin x_2(\tau+t)] \\ & +E[\sin x_1(\tau) \cos x_2(\tau) \cos x_1(\tau+t) \sin x_2(\tau+t)] \\ & +E[\cos x_1(\tau) \sin x_2(\tau) \sin x_1(\tau+t) \cos x_2(\tau+t)] \end{aligned}$$

If $x_1(t)$ and $x_2(t)$ are statistically independent, the first line (for example) of the above equation for $R(t)$ may be written

$$E[\sin x_1(\tau) \sin x_1(\tau+t)] \,.\, E[\cos x_2(\tau) \cos x_2(\tau+t)]$$

If it is further assumed that $\sin x_1(t)$ and $\cos x_1(t)$ are uncorrelated (see Appendix 4), then the last two lines vanish and $R(t)$ is now given by

$$\begin{aligned} R(t) = & E[\sin x_1(\tau) \sin x_1(\tau+t)].E[\cos x_2(\tau) \cos x_2(\tau+t)] \\ & +E[\cos x_1(\tau) \cos x_1(\tau+t)].E[\sin x_2(\tau) \sin x_2(\tau+t)] \end{aligned}$$

This equation may now be transformed giving results in terms of double sided power spectra. Writing $S_1(f)$ and $C_1(f)$ for the power spectra of $\sin x_1(t)$ and $\cos x_1(t)$ respectively, the spectrum of $V(t)$ is given by

$$S_{1+2}(f) = S_1(f)*C_2(f)+C_1(f)_2*S_2(f) \qquad \text{(A3.1)}$$

where $*$ represents the process of convolution, the transform of the multiplication operation.

i.e. $$S(f)*C(f) = \int_{-\infty}^{\infty} S(\alpha)\, C(f-\alpha)\, d\alpha$$

It may also be shown that

$$C_{1+2}(f) = C_1(f)*C_2(f)+S_1(f)*S_2(f) \qquad \text{(A3.2)}$$

In the particular case of an angle modulated wave of frequency ω, $x_1(t)$ is put equal to ωt and $x_2(t)$ to the angle modulation $\theta(t)$.

Thus, the spectrum $W(\alpha)$ of the waveform $\sqrt{2}\sin(\omega t+\theta(t))$ which has unit total power is

$$W(\alpha) = 2(S_\omega(\alpha)*C_\theta(\alpha)+C_\omega(\alpha)*S_\theta(\alpha)$$

$$\text{Now } S_\omega(\alpha) = C_\omega(\alpha) = \tfrac{1}{4}\,\delta(\alpha-f_0)+\tfrac{1}{4}\,\delta(\alpha+f_0)$$

where $\delta(\alpha)$ is the Dirac delta function. Since convolution of a function $G(\alpha)$ with the delta function $\delta(\alpha-f_0)$ translates the function $G(\alpha)$ to $G(\alpha-f_0)$, we have

$$W(\alpha) = \tfrac{1}{2}\,(C_\theta(\alpha-f_0)+S_\theta(\alpha-f_0)+C_\theta(\alpha+f_0)+S_\theta(\alpha+f_0)) \qquad \text{(A.3.3)}$$

It is convenient to express the power spectrum of a modulated carrier in terms of a parameter relative to the carrier frequency and hence the contribution of the spectrum due to the power in the negative frequency plane must be added to that in the positive.

Putting $f = \alpha-f_0$, the spectrum relative to the frequency of the carrier f_0, $W_y(f)$ is given by

$$W_y(f) = C_\theta(f)+S_\theta(f)+C_\theta(f+2f_0)+S_\theta(f+2f_0)$$

If f_0 is a microwave carrier whose frequency is very large compared with the values of f for which $C_\theta(f)$ and $S_\theta(f)$ are significant, the contribution of $C_\theta(f+2f_0)$ and $S_\theta(f+2f_0)$ may be neglected.

$$W_y(f) = C_\theta(f)+S_\theta(f)$$

Thus the spectrum of a phase modulated wave about the carrier frequency is the sum of the spectrum of the cosine of the modulating waveform and the spectrum of the sine of the modulating waveform.

Further, a function of the form $\sqrt{2}\sin(\omega_0 t+\theta(t)+\phi(t))$ has a spectrum which, from equation A3.3, may be expressed as follows

$$W(\alpha) = \tfrac{1}{2}\,(C_{\theta+\phi}(\alpha-f_0)+S_{\theta+\phi}(\alpha-f_0)+C_{\theta+\phi}(\alpha+f_0)+S_{\theta+\phi}(\alpha+f_0)) \qquad \text{(A3.4)}$$

Again, if f_0 is large,

$$W_y(f) = C_{\theta+\phi}(f) + S_{\theta+\phi}(f)$$

From equations A3.1 and A3.2

$$W_y(f) = C_\theta(f) * C_\phi(f) + S_\theta(f) * S_\phi(f)$$
$$+ S_\theta(f) * C_\phi(f) + C_\theta(f) * S_\phi(f)$$
$$W_y(f) = (S_\theta(f) + C_\theta(f)) * (S_\phi(f) + C_\phi(f)) \qquad \text{(A3.5)}$$
$$W_y(f) = W_{y\theta}(f) * W_{y\phi}(f)$$

Equation A3.5 shows that the spectrum about a carrier which is modulated by the sum of two statistically independent phase disturbances is the convolution of the spectra about two individual carriers modulated by each phase disturbance separately.

Since, from equation A3.4,

$$W(\alpha) = \left(\frac{\delta}{2}(\alpha - f_0) + \frac{\delta}{2}(\alpha + f_0)\right) * (C_{\theta+\phi}(\alpha) + S_{\theta+\phi}(\alpha)) \qquad \text{(A3.6)}$$

then, if f_0 cannot be considered large,

$$2W(\alpha) = W_y(\alpha - f_0) + W_y(\alpha + f_0)$$

CORRELATION BETWEEN Cos $x(t)$ AND Sin $x(t)$

The assumption that cos $x(t)$ and sin $x(t)$ are uncorrelated crops up a number of times in this book. The assumption is verified in this appendix for a particular class of function for $x(t)$. Here it is taken that $x(t)$ can be expressed as the sum of a number of sinusoids (i.e. can be Fourier transformed) which covers all the usual functions encountered in this aspect of communications.

Thus it is sufficient to demonstrate that the functions,

$\cos(\beta \sin \omega t)$ and $\sin(\beta \sin \omega t)$ are uncorrelated.

Expressing these functions in terms of Bessel functions,

$$\cos(\beta \sin \omega t) = J_0(\beta) + 2 \sum_{n=1}^{\infty} J_{2n}(\beta) \cos 2n\omega t$$

$$\sin(\beta \sin \omega t) = 2 \sum_{n=0}^{\infty} J_{2n+1}(\beta) \sin(2n+1)\omega t$$

where the J_x are Bessel coefficients.

The correlation between the functions may therefore be expressed as

$$\lim_{T\to\infty} \frac{1}{2T} \int_{-T}^{T} \cos 2n\omega(\tau+t) \sin(2m+1)\omega\tau \, d\tau$$

$$= \lim_{T\to\infty} \frac{1}{4T} \int_{-T}^{T} \{\sin[2n\omega(\tau+t)+(2m+1)\omega\tau] + \sin[(2m+1)\omega\tau - 2n\omega(\tau+t)]\} \, d\tau$$

$$= \lim_{T\to\infty} \frac{1}{4T} \int_{-T}^{T} \left\{\sin\left[2\left(m+n+\frac{1}{2}\right)\omega\tau + 2n\omega t\right] + \sin\left[2\left(m-n+\frac{1}{2}\right)\omega\tau - 2n\omega t\right]\right\} d\tau$$

$$= 0$$

Appendix 5

SOME USEFUL RELATIONSHIPS FOR USE WITH THE CHARACTERISTIC FUNCTION METHOD

N. B. A more comprehensive list of formulae of this type may be found in the Appendices of Reference 1 of Chapter 1

Gamma function

$$\Gamma(z) = \int_0^\infty e^{-t}t^{z-1}\,dt$$

$$\Gamma(\tfrac{1}{2}) = \sqrt{\pi}$$

$$\Gamma(-n) = \infty \qquad n \quad +\text{ve integer.}$$

$$\Gamma(z)\Gamma(1-z) = \frac{\pi}{\sin \pi z}$$

$$(x)\Gamma(x) = \Gamma(x+1)$$

Bessel Function

$$I_v(jz) = j^v I_v(z)$$

$$j^v I_v(z) = J_v(jz)$$

Confluent Hypergeometric Function

$${}_1F_1(\alpha, \beta, x) = 1+\frac{\alpha x}{\beta 1!}+\frac{\alpha(\alpha+1)}{\beta(\beta+1)}\frac{x^2}{2!}\cdots\frac{(\alpha)_n x}{(\beta)_n n!}\cdots$$

where $(\alpha)_n = \alpha(\alpha+1)\ldots(\alpha+n-1)$, and $(\alpha)_0 = 1$.

$${}_1F_1(\alpha; \beta; -x) \approx \frac{\Gamma(\beta)}{\Gamma(\beta-\alpha)}.x^{-\alpha}\left\{1+\frac{\alpha(\alpha-\beta+1)}{x1!} + \frac{\alpha(\alpha+1)(\alpha-\beta+1)(\alpha-\beta+2)}{x^2 2!}+\ldots\right\}$$

$$(+\text{ve } x)$$

$${}_1F_1(\alpha; \beta; x) = e^x\,{}_1F_1(\beta-\alpha, \beta, -x)$$

$${}_1F_1(\alpha; \beta; -x) \to 1 \text{ as } x \to 0$$

$${}_1F_1(\alpha; \beta; -x) \to \frac{\Gamma(\beta)}{\Gamma(\beta-\alpha).x^\alpha} \text{ as } x \to \infty$$

$${}_2F_1(\alpha, \beta; \gamma; 1) = \frac{\Gamma(\gamma).\Gamma(\gamma-\alpha-\beta)}{\Gamma(\gamma-\alpha)\Gamma(\gamma-\beta)}$$

Solution of the following integrals may be found in Appendix 3 of Reference 11 of Chapter 5.

$$\int_c J_v(az)\mathrm{e}^{-q^2z^2}z^{\mu-1}\,\mathrm{d}z \qquad \int_c J_\alpha(az)\,J_\beta(bz)\,\mathrm{d}z/z^\gamma$$

$$\int_c z^\mu\,\mathrm{e}^{\mathrm{j}zb_0-c^2z^2}\,\mathrm{d}z \qquad \int_c \mathrm{e}^{-c^2z^2}\,z^{2\mu-1}\,\mathrm{d}z$$

INDEX